OPTIMIZATION
USING
LINEAR
PROGRAMMING

OPTIMIZATION USING LINEAR PROGRAMMING

An Introduction

A.J. Meitei, PhD
Veena Jain, PhD

MERCURY LEARNING AND INFORMATION
Dulles, Virginia
Boston, Massachusetts
New Delhi

Publisher: David Pallai
MERCURY LEARNING AND INFORMATION
22841 Quicksilver Drive
Dulles, VA 20166
info@merclearning.com
www.merclearning.com
(800) 232-0223

A.J. Meitei and Veena Jain. *Optimization Using Linear Programming*.
ISBN: 978-1-68392-347-3

The publisher recognizes and respects all marks used by companies, manufacturers, and developers as a means to distinguish their products. All brand names and product names mentioned in this book are trademarks or service marks of their respective companies. Any omission or misuse (of any kind) of service marks or trademarks, etc. is not an attempt to infringe on the property of others.

Library of Congress Control Number: 2018964994

192021321   Printed on acid-free paper in the United States of America.

Our titles are available for adoption, license, or bulk purchase by institutions, corporations, etc. For additional information, please contact the Customer Service Dept. at (800) 232-0223(toll free). Digital versions of our titles are available at: www.academiccourseware.com and other electronic vendors.

CONTENTS

PREFACE

This book on linear programming and game theory has been jointly written by Dr. A. J. Meitei and Dr. Veena Jain with an aim to meet the needs of the students of mathematics, commerce, economics, management studies, and other allied disciplines or courses. The explanation and presentation of every topic in the book have been made as simple and user-friendly as possible. Complex mathematics involved in various theorems and procedures has been avoided, and all explanations are given in simplified and systematic forms so that even non-mathematical students or those who know only basic mathematics can easily and conveniently read the book. The main emphasis is on the solution of various types of linear programming problems by using different kinds of software. Use of software in solving mathematical problems has been an integral part of syllabi these days. Keeping this in mind, the solution of problems using the MS-Excel Solver add-in and the external Jenson add-in have been discussed in all chapters of this book. We explain step by step the procedure of how the add-ins can be used to solve linear programming problems. In addition to MS-Excel, solutions of LPPs by *Mathematica*, MATLAB, WinQSB, and LINDO have also been explained in the Appendix.

Exercises are given at the end of each chapter so that students can practice a variety of problems. In order to make it easy for students to follow along, all of the materials related to various topics are arranged in a systematic way. All the definitions, theorems, and procedures for solving problems and all cases related to the various topics are discussed clearly in simple language.

The book is divided into nine chapters. At the beginning, Chapter 1 discusses the basic concepts of algebra that include vectors, matrices, operations on matrices and other related methods like the Gauss-Jordan method, solutions of simultaneous linear equations, convex sets, and so forth. The use of MS-Excel in algebraic computations is also explained with relevant examples. All of these concepts are used in developing and understanding the solution procedure for solving a Linear Programming Problem (LPP), so it was essential to incorporate them in the book as a separate chapter. Chapter 2 explains each definition along with the formulation and graphical method for the solution of a linear programming problem. Some important definitions and theorems related to the solution of linear programming problems have also been incorporated. Also, the use of MS-Excel for plotting graphs and finding the solution of an LPP is thoroughly explained with examples. Chapter 3 focuses on solving linear programming problems by the simplex method with the help of its canonical form in a slightly different manner, which has been explained by very few authors. In Chapter 4, the M-Charnes and two Phase-methods are included, in which the manual solution procedure and the solution by using Excel the Solver and the Jensen add-in have also been discussed in detail. In addition, a detailed discussion of various special LPP using both Excel Solver and simplex tables is included in the chapter. The concept of duality with its related theorems and importance is the main topic explained in detail in Chapter 5. In Chapter 6, a sensitivity analysis is carried out in a linear programming problem by considering all possible changes in the parameters and structure of the LPP. Chapters 7 and 8 are on transportation, transshipment, and assignment problems. In these chapters the definition and procedure for solving these types of problems are discussed at length. Chapter 9 is on game theory, where the solution of game problems using different techniques is explained and the use of Gambit Software for finding solutions is discussed as well. Suggestions for further enhancement are welcome.

Dr. A. J. Meitei

Dr. Veena Jain

CHAPTER 1

BASICS OF LINEAR ALGEBRA USING MS-EXCEL

1.1. Vectors

An arrangement of elements either in a row or in a column is called a vector and is usually denoted by lowercase bold letters like *a, b, c,* and so on.

$a = (a_1, a_2, a_3)$ is a row vector of three elements, and a_i, where $i = 1, 2, 3$, is said to be the ith element of *a*. Similarly, $a = \begin{pmatrix} 5 \\ 3 \end{pmatrix}$ is a column vector with two elements.

Geometrically, any vector $a = (a_1, a_2)$ can be considered as a point in a 2-dimensional space. In general a vector $a = (a_1, a_2, a_3, ..., a_n)$ can be considered as a point in an *n*-dimensional space.

Equality of two vectors: Two *n*-component vectors $a = (a_1, a_2, a_3, ..., a_n)$ and $b = (b_1, b_2, b_3, ..., b_n)$ are said to be equal if $a_i = b_i$ *for all* $i = 1, 2, 3, ..., n$. It should also be noted that if *a = b* then *b = a.*

Addition of vectors: Let $a_1 = (2, 4, 6, 9)$ and $a_2 = (1, 4, 5, 2)$ be any two vectors from a 4-dimensional real space. Then the addition of a_1 and a_2, denoted by $a_1 + a_2$, is given as follows:

$a_1 + a_2 = (2 + 1, 4 + 4, 6 + 5, 9 + 2) = (3, 8, 11, 11)$. To solve this using Excel, we can use the command for matrix addition. This operation shown in the screenshot will be explained later in matrix addition.

Dot or Inner Product of Vectors: The inner or dot product of two vectors will be defined only if the vectors have the same number of components. Let a_1 and a_2 be any two real vectors from an *n*-dimensional real space. Then the inner or dot product of a_1 and a_2 is given by,

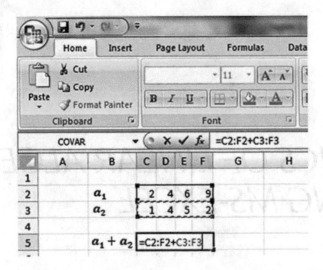

Fig. 1.1

$$a_1 \cdot a_2 = \sum_{i=1}^{n} a_{1i}\, a_{2i}$$

Let $a_1 = (2\ 4\ 6)$ and $a_2 = (1\ 4\ 5)$, and then $a_1 \cdot a_2 = 2 \times 1 + 4 \times 4 + 6 \times 5 = 48$. It is also to be noted that the inner product of any two vectors is always a scalar.

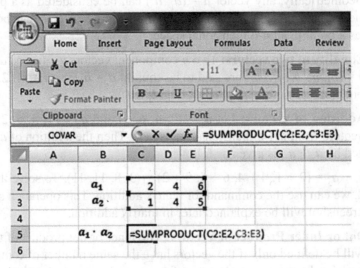

Fig. 1.2

In Excel the **SUMPRODUCT** function can be used to find the dot product of any two vectors of the same dimension.

Zero Vector: A vector whose elements are all zero is called a zero vector, and it is usually denoted by **0**. This vector is also referred to as the origin. In the *XY* plane, (0 0) is a zero vector with two components.

Unit Vector: A vector denoted by e_i whose i^{th} component is 1 and all the remaining components are zero is called the ith unit vector. For a 3-dimensional space there are three unit vectors, namely $e_1 = (1, 0, 0)$, $e_2 = (0, 1, 0)$, and $e_3 = (0, 0, 1)$.

Sum Vector: A vector whose elements are all 1 is called a sum vector and is denoted by **1**; that is, $\mathbf{1} = (1, 1, \dots, 1)$.

Euclidean Space: This space, sometimes called Cartesian space or simply *n* space, is the space of all *n*-tuples of real numbers $(x_1, x_2, \dots x_n)$ and is generally denoted by R^n *or* E^n.

Matrix: A rectangular arrangement of numbers into rows and columns is called a matrix and is always enclosed in either brackets [] or parentheses (). If the matrix has *m* rows and *n* columns, it is called an *m* × *n* matrix (read as "*m*" by "*n*"). *m* × *n* is called the dimension of the matrix. It is usually denoted by capital boldface letters, such as *A, B, C,* and so forth. A matrix has no numerical value, and the numbers in the matrix are called elements of the matrix. A double subscript is used to denote the location of the element in the matrix, where the first subscript indicates the row number and the second subscript indicates the column number. For example:

$$A = \begin{pmatrix} a_{11} & a_{12} \\ a_{21} & a_{22} \end{pmatrix}$$ is a 2 × 2 matrix or 2 by 2 matrix, and a_{ij} is the element

in the ith row and j^{th} column of the given matrix where $i = 1, 2$ and $j = 1, 2$.

Square Matrix: A matrix whose number of rows are equal to the number of columns is called a square matrix. For example, $C = \begin{pmatrix} 2 & 8 \\ 9 & 5 \end{pmatrix}$ is a 2 × 2 square matrix.

Zero Matrix: If each element in a matrix is zero, then the matrix is said to be a zero or null matrix; $C = \begin{pmatrix} 0 & 0 & 0 \\ 0 & 0 & 0 \end{pmatrix}$ is a 2 × 3 zero matrix. A null matrix need not be a square matrix.

Identity Matrix: A square matrix denoted by *I*, in which all diagonal elements are one and the other elements are zero, is called an identity matrix.

An $m \times m$ identity matrix is denoted by I_m. It should also be noted that the multiplication of an identity matrix with any other matrix is the matrix itself, that is, $AI_m = I_m A = A$, where A is any $m \times m$ matrix.

Determinant: It is a number which is associated with every square matrix. The determinant of the nth order matrix A denoted by $|A|$ is computed as follows:

$$|A| = \sum (\pm) a_{1i} \, a_{2j} \ldots \ldots \ldots a_{nr}$$

where the sum is taken over all permutations of the second subscript. A plus sign is assigned to even permutations and a minus sign to odd permutations.

Consider a third-order matrix $A = \|a_{ij}\| = \begin{bmatrix} a_{11} & a_{12} & a_{13} \\ a_{21} & a_{22} & a_{23} \\ a_{31} & a_{32} & a_{33} \end{bmatrix}$

Then $|A| = a_{11}a_{22}a_{33} - a_{12}a_{21}a_{33} + a_{12}a_{23}a_{31} - a_{13}a_{22}a_{31} + a_{13}a_{21}a_{32} - a_{11}a_{23}a_{32}$

In Excel, we can use the **MDETERM** function to find the determinant of any square matrix as follows:

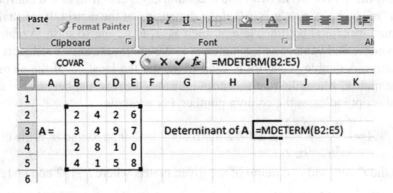

Fig. 1.3

Singular Matrix: A square matrix B is said to be a singular matrix if its determinant is zero; otherwise, it is non-singular. For example:

$$A = \begin{pmatrix} 2 & 4 & 2 & 6 \\ 3 & 4 & 9 & 7 \\ 2 & 8 & 1 & 0 \\ 4 & 1 & 5 & 8 \end{pmatrix} \text{ is a non-singular matrix, as } |A| = 668 \neq 0.$$

$B = \begin{pmatrix} 2 & 4 \\ 5 & 10 \end{pmatrix}$ is a singular matrix, as $|B| = 0$.

Triangular Matrix: Any square matrix is said to be an upper triangular matrix if all the entries below the main diagonal are zeros. Similarly, any square matrix is called a lower triangular matrix if entries above the main diagonal of the matrix are zeros.

For example, $B = \begin{pmatrix} 2 & 4 & 8 \\ 0 & 4 & 7 \\ 0 & 0 & 5 \end{pmatrix}$ is an upper triangular matrix, and $C = \begin{pmatrix} 2 & 0 & 0 \\ 1 & 4 & 0 \\ 0 & 1 & 5 \end{pmatrix}$

is a lower triangular matrix.

Multiplication of a Matrix by a Scalar: Let A be an $m \times n$ matrix and k be any scalar. Then $B = kA$ is an $m \times n$ matrix whose every element is k times the corresponding element of A.

Let $A = \begin{pmatrix} 2 & 4 & 2 & 6 \\ 3 & 4 & 9 & 7 \\ 2 & 8 & 1 & 0 \\ 4 & 1 & 5 & 8 \end{pmatrix}$ and $k = 4$. Then $B = kA = \begin{pmatrix} 8 & 16 & 8 & 24 \\ 12 & 16 & 36 & 28 \\ 8 & 32 & 4 & 0 \\ 16 & 4 & 20 & 32 \end{pmatrix}$

To perform this calculation in Excel, select the output space for B, then multiply the matrix A by k as follows, and finally press **Ctrl, Shift, and Enter** simultaneously.

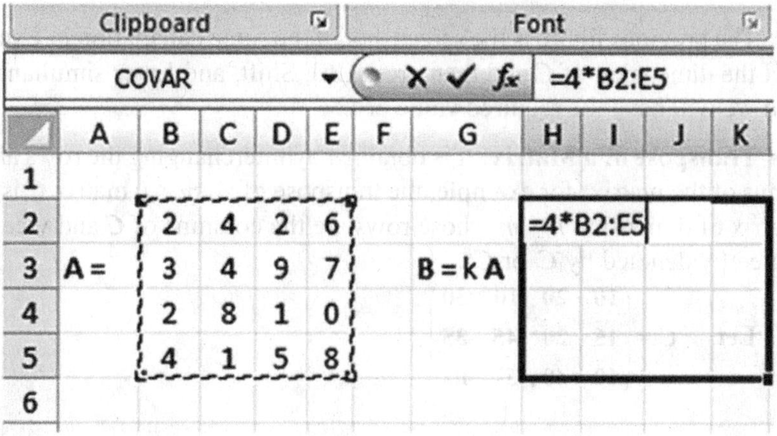

Fig. 1.4

Addition of Matrices: The addition of two matrices is defined only if they are of the same dimension. The previous matrices A and B are of the same dimension, 4×4, and their addition $C = A + B$ is another matrix of same dimension whose elements are the sum of the corresponding elements of the matrices A and B.

$$C = \begin{pmatrix} 10 & 20 & 10 & 30 \\ 15 & 20 & 45 & 35 \\ 10 & 40 & 5 & 0 \\ 20 & 5 & 25 & 40 \end{pmatrix}$$

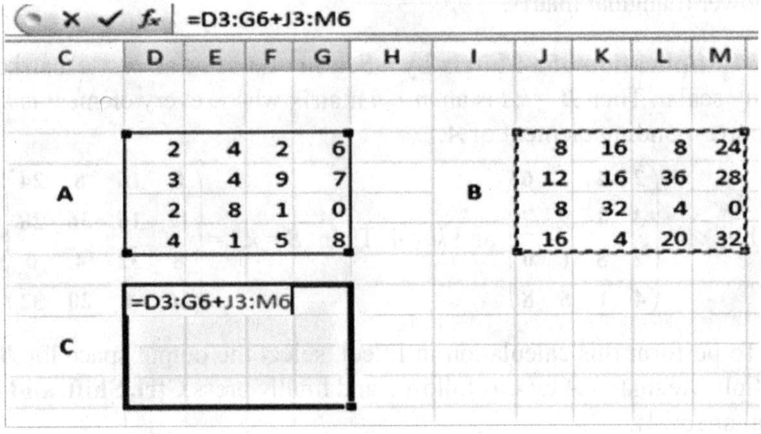

Fig. 1.5

The previous figure is the screenshot of the same calculation in Excel. Select the dimension of C and then press Ctrl, Shift, and Enter simultaneously, and we will have the required value of C.

Transpose of a Matrix: It is obtained by interchanging the rows and columns of the matrix; for example, the transpose of an $m \times n$ matrix C is a new matrix of dimension $n \times m$ whose rows are the columns of C and vice versa, generally denoted by C' or C^T.

$$\text{Let} \quad C = \begin{pmatrix} 10 & 20 & 10 & 30 \\ 15 & 20 & 45 & 35 \\ 10 & 40 & 5 & 0 \end{pmatrix}$$

Then $\qquad C' = \begin{pmatrix} 10 & 15 & 10 \\ 20 & 20 & 40 \\ 10 & 45 & 5 \\ 30 & 35 & 0 \end{pmatrix}$

In Excel, we can use the **TRANSPOSE** function to find the transpose of a given matrix.

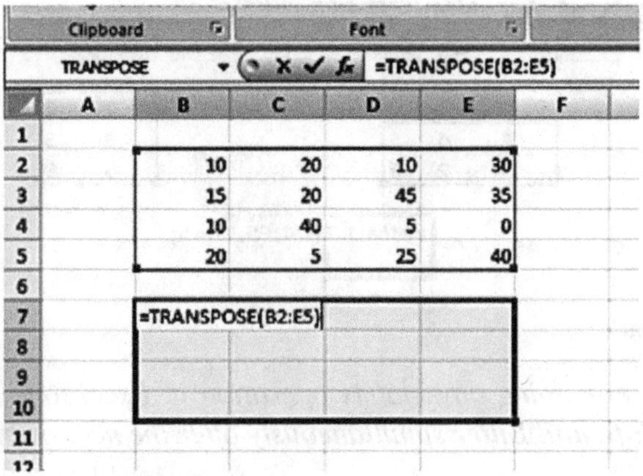

	A	B	C	D	E	F
1						
2		10	20	10	30	
3		15	20	45	35	
4		10	40	5	0	
5		20	5	25	40	
6						
7		=TRANSPOSE(B2:E5)				
8						
9						
10						
11						

Fig. 1.6

Matrix Multiplication: The multiplication of any two matrices is defined only if the number of columns of the first matrix is equal to the number of rows of the second matrix. Let A be an $m \times n$ and B be an $n \times p$ matrix. Then their product is another matrix $C \, (= AB)$ of order $m \times p$ with:

$$c_{ij} = \sum_{k=1}^{n} a_{ik} b_{kj} \text{ for } i = 1, 2, 3, \ldots, m \text{ and } j = 1, 2, 3, \ldots, p$$

Example: $A = \begin{pmatrix} 2 & 4 & 5 \\ 3 & 2 & 3 \end{pmatrix}$ and $B = \begin{pmatrix} 1 & 2 \\ 3 & 5 \\ 6 & 9 \end{pmatrix}$

Then

$$C = AB = \begin{pmatrix} 2 & 4 & 5 \\ 3 & 2 & 3 \end{pmatrix} \begin{pmatrix} 1 & 2 \\ 3 & 5 \\ 6 & 9 \end{pmatrix} = \begin{pmatrix} 2*1+4*3+5*6 & 2*2+4*5+5*9 \\ 3*1+2*3+3*6 & 3*2+2*5+3*9 \end{pmatrix}$$

$$= \begin{pmatrix} 44 & 69 \\ 27 & 43 \end{pmatrix}$$

In Excel the **MMULT** function can be used for matrix multiplication.

Step 1. Select the dimension of the matrix C in the output space.

Step 2. Type the command MMULT.

Step 3. Select the two matrices as shown in the following figure.

Step 4. Finally, press Ctrl, Shift, and Enter simultaneously.

				fx	=MMULT(D4:F5,J3:K5)					
C	**D**	**E**	**F**	**G**	**H**	**I**	**J**	**K**		
							1	2		
A		2	4	5		B	3	5		
		3	2	3			6	9		
C	AB	=MMULT(D4:F5,J3:K5)								

Fig. 1.7

*Remark: For doing any matrix operation in Excel, one should always press **Ctrl, Shift, and Enter** simultaneously after the necessary inputs.*

Vector Space: A vector space is a space consisting of a collection of vectors which are closed under the operation of addition and multiplication by a scalar; that is, if vectors a, b are in a collection, then $a + b$ and ka will also be in the collection, where k is a scalar quantity.

Rank: The rank of any matrix A, written as $r(A)$, is the maximum number of linearly independent columns in A, or it is the order of the largest non-vanishing minor (determinant of the square submatrix) in A. The rank of a matrix is always unique, since the row rank is always equal to the column rank; that is, the maximum number of linearly independent columns in a matrix is always equal to the maximum number of linearly independent rows.

For example, $A = \begin{bmatrix} 1 & 1 & 1 \\ 1 & 1 & 1 \\ 1 & 1 & 1 \end{bmatrix}$ has rank 1 since $|A| = 0$ and every minor of order 2 also vanishes.

Note: The rank of a matrix A will be equal to the dimension of the largest square sub-matrix of A which is non-singular.

Example 1.1. *Show that the rank of* $\begin{bmatrix} 0 & 0 & 0 \\ 0 & 0 & 0 \\ 0 & 0 & 0 \end{bmatrix}$ *is zero.*

Solution: We cannot identify any sub-matrix of the given matrix which is non-singular, and hence the rank of the matrix is zero.

Example 1.2. *Show that the rank of* $\begin{pmatrix} 2 & 5 & 6 \\ 1 & 8 & 3 \\ 1 & 3 & 1 \end{pmatrix}$ *is 3.*

Solution: The determinant of the largest sub-matrix of the given matrix, which is different from zero, is the matrix itself. Hence the rank of the given matrix is 3.

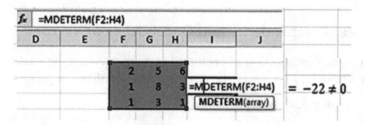

Fig. 1.8

Example 1.3. *Show that the rank of* $\begin{pmatrix} 2 & 5 & 6 \\ 1 & 8 & 3 \\ 2 & 16 & 6 \end{pmatrix}$ *is 2.*

Solution: The determinant of the largest order sub-matrix of the given matrix, which is different from zero, is of dimension 2×2. Hence the rank of the given matrix is 2.

Matrix Inverse: An $n \times n$ square matrix B is said to be the inverse of another $n \times n$ non-singular square matrix A if $BA = I$, where I is the identity matrix of the same dimension. The inverse of matrix A is usually denoted by A^{-1}.

In Excel we can use the **MINVERSE** function to find the inverse of any square matrix.

× ✓ f_x	=MINVERSE(D2:F4)										
C	D	E	F	G	H	I	J	K			
	2	4	6		=MINVERSE(D2:F4)				0.04	-0.6	1.5
A	1	8	3		A^{-1}			=	-0.08	0.17	0
	1	3	1						0.21	0.08	-0.5

Fig. 1.9

Example 1.4. *Use the* **MINVERSE** *function to find the inverse of the following matrix:*

$$A = \begin{pmatrix} 2 & 5 & 6 \\ 1 & 8 & 3 \\ 1 & 3 & 1 \end{pmatrix}$$

Solution: The following is the screenshot of the Excel calculation of the inverse of matrix A.

1.2. Linear Independence and Dependence of Vectors

Linear combination of vectors: Let $a_1, a_2, a_3, \ldots, a_k$ be a set of k vectors from R^n and $\lambda_1, \lambda_2, \lambda_3, \ldots, \lambda_k$ be any k scalars, and then the vector

$$c = \lambda_1 a_1 + \lambda_2 a_2 + \lambda_3 a_3 + \ldots + \lambda_k a_k$$

is known as a linear combination of vectors $a_1, a_2, a_3, \ldots, a_k$.

Linearly dependent vectors: A set of vectors $a_1, a_2, a_3, \ldots, a_k$ from R^n is said to be linearly dependent if there exist scalars $\lambda_1, \lambda_2, \lambda_3, \ldots, \lambda_k$ that are not all zero, such that

$$\lambda_1 a_1 + \lambda_2 a_2 + \lambda_3 a_3 + \ldots + \lambda_k a_k = 0$$

And if the previous equation holds only when all λ_i $(i = 1, 2, 3, \ldots, k)$ are zero, then the vectors are said to be linearly independent.

Note: To check the linear independence of vectors, we can write the linear combination of the given vectors as a system of linear equations of the form $\lambda A = 0$ and solve for λ. If the solution contains at least one $\lambda i \neq 0$, then the set of vectors is linearly dependent; otherwise, it is linearly independent.

Example 1.5. *The vectors $a_1 = (2, 6)$ and $a_2 = (4,12)$ are linearly dependent vectors, as we can find $\lambda_1 = 2$ and $\lambda_2 = -1$, for which $\lambda_1 a_1 + \lambda_2 a_2 = 0$.*

Example 1.6. *The vectors $e_1 = (1,0,0)$, $e_2 = (0,1,0)$, and $e_3 = (0,0,1)$ are linearly independent. We have,*

$$\lambda_1 e_1 + \lambda_2 e_2 + \lambda_3 e_3 = 0$$
$$\Rightarrow \quad \lambda_1 (1,0,0) + \lambda_2 (0,1,0) + \lambda_3 (0,0,1) = (0,0,0)$$
$$\Rightarrow \quad (\lambda_1, \lambda_2, \lambda_3) = (0,0,0)$$
$$\Rightarrow \quad \lambda_1 = 0, \lambda_2 = 0 \quad and \quad \lambda_3 = 0$$

Hence, the set of unit vectors is always linearly independent.

Notes:

(i) A *null vector is not linearly independent of any other vector or set of vectors.*

(ii) *If a set of vectors is linearly independent, then any subset of these vectors is also linearly independent.*

(iii) *If any set of vectors is linearly dependent, then any larger set of vectors containing these vectors is also linearly dependent.*

(iv) *Any vector x is said to be linearly dependent on a set of vectors $x_1, x_2,...,$ x_k if x can be written as a linear combination of the set of vectors.*

(v) *If $x_1, x_2,... , x_k$ is a given set of vectors from R^n and there exists at least one subset of $r < k$ vectors which are linearly independent but no subset containing $(r + 1)$ vectors is linearly independent, then r is the maximum number of linearly independent vectors in the given set. Given this subset of r linearly independent vectors in the set, any other vector in the set can be written as a linear combination of these r vectors.*

(vi) *A set of vectors $b_1, b_2 ..., b_k$ from R^n where $k \geq 2$ is linearly independent if and only if one of these vectors can be written as a linear combination of the others.*

Spanning Set: A set of vectors $a_1, a_2, ..., a_k$ $(k \geq 2)$ from R^n is said to span or generate R^n if every vector in R^n can be written as a linear combination of the given set of vectors. The vectors in the spanning set must be linearly independent.

Basis: A basis for R^n is a subset of linearly independent vectors from R^n which spans the entire space.

Notes:

(i) *There exist an infinite number of bases in R^n.*

(ii) *A set of unit vectors will always form a basis, since it is linearly independent and any vector in the space can be written as a linear combination of unit vectors.*

(iii) *The basis formed by the set of unit vectors is called a standard basis.*

Theorem 1.1. The set of unit vectors forms a basis.

Proof. Let e^i ($i = 1, 2, 3, \ldots, n$) denote the set of n unit vectors from $\boldsymbol{R}^n$. Then we will have to show that $e_i's$ is linearly independent and any vector of $\boldsymbol{R}^n$ should be able to be expressed as a linear combination of these unit vectors.

Let $\lambda_i's$ be n scalars, and then we have

$$\sum_{i=1}^{n} \lambda_i e_i = 0$$

$$\Rightarrow \lambda_1(1,0, \ldots, 0) + \lambda_2(0,1, \ldots, 0) + \ldots + \lambda_n(0,0, \ldots, 1) = (0,0, \ldots 0)$$

$$\Rightarrow (\lambda_1, \lambda_2, \ldots, \lambda_n) = (0,0, \ldots 0)$$

This implies e_i ($i = 1,2,3, \ldots, n$) is linearly independent.

Let $x = (x_1, x_2, \ldots, x_n)$ be any other vector of $\boldsymbol{R}^n$ different from $e_i's$. Then we can express the vector x as a linear combination of the n unit vectors as follows:

$$x = (x_1, x_2, \ldots, x_n) = x_1 e_1 + x_2 e_2 + \ldots + x_n e_n$$

Since x is any vector from $\boldsymbol{R}^n$ different from $e_i(i = 1,2,3, \ldots, n)$, every vector of $\boldsymbol{R}^n$ can be expressed as a linear combination of these unit vectors. Hence, the set of unit vectors always forms a basis for the given space.

1.3. Solution to a System of Simultaneous Linear Equations

Consider a system of m simultaneous linear equations in n unknowns of the form

$$a_{11}x_1 + a_{12}x_2 + \ldots a_{1n}x_n = b_1$$
$$a_{21}x_1 + a_{22}x_2 + \ldots a_{2n}x_n = b_2$$
$$\vdots \qquad\qquad \vdots$$
$$a_{m1}x_1 + a_{m2}x_2 + \ldots + a_{mn}x_n = b_m$$

...(1.1)

In the set of equations (1.1), $x_1, x_2, \ldots, x_n$ are the unknown decision variables, and $a_{ij}'s$ and $b_i's$ are constants where $i = 1,2, \ldots m$, and $j = 1,2 \ldots n$.

A solution to (1.1) is the set of values of the unknown variables $x_j's$, $j = 1 \ldots \ldots n$ which satisfies all the m equations of (1.1).

For understanding linear programming we need to understand the properties of solutions to the linear system of equations. Keeping this in mind, we will devote some effort to studying such systems. The matrix representation of the set of equations (1.1) can be written as,

$$Ax = b$$

Where $A = \begin{pmatrix} a_{11} a_{12} & \cdots & a_{1n} \\ a_{21} a_{22} & \cdots & a_{2n} \\ \vdots & \vdots & \vdots \\ a_{m1} a_{m2} & \cdots & a_{mn} \end{pmatrix}_{m \times n}$, $x = \begin{pmatrix} x_1 \\ x_2 \\ \vdots \\ x_n \end{pmatrix}_{n \times 1}$ & $b = \begin{pmatrix} b_1 \\ b_2 \\ \vdots \\ b_m \end{pmatrix}_{m \times 1}$

The system of simultaneous linear equations may have either no solution or at least one solution. To determine the conditions under which the system has at least one solution, form an augmented matrix $A|b = (A,b)$ of the order $m \times (n + 1)$ containing the whole matrix A and the vector b. We always have $r(A|b) \geq r(A)$ since every minor of A also appears in $A|b$. Now

(i) If $r(A|b) > r(A)$, then no solution exists to the given system of linear equations.

(ii) If $r(A|b) = r(A)$, then there exists at least one solution.

If the system of equations has at least one solution, then the given set of equations is called consistent; otherwise, it is said to be inconsistent. Further:

(i) If $m = n$ and $r(A|b) = r(A) = m$, then there exists a unique solution to (1.1).

(ii) If $m < n$ and $r(A|b) = r(A) = k < m$, then $m - k$ of the equations are redundant. Any solution which satisfies k equations will also satisfy $m - k$ of the equations.

(iii) If $m < n$ and $r(A|b) = r(A) = m$, then there exist an infinite number of solutions to the given system.

Basic Solution: Given a system of m simultaneous linear equations in n unknowns ($m < n$),

$$A_{m \times m} x_{n \times 1} = b_{m \times 1}, (m < n)$$

Let $\mathbf{B}_{m \times m}$ be any $m \times m$ non-singular sub-matrix of $A_{m \times n}$. Then, the solution obtained by setting the $(n - m)$ variables not associated with the columns of $\mathbf{B}_{m \times m}$ equal to zero is called a basic solution to the given system of equations.

Let the set of m variables associated with the columns of $\mathbf{B}_{m \times m}$ be denoted by x_B and the remaining $(n - m)$ variables by $x_N B (= 0)$, and then

$$B_{m \times m} x_B = b_{m \times 1} \Rightarrow x_B = B_{m \times m}^{-1} b_{m \times 1}$$

is the basic solution for the given system of equations.

Notes:

(i) If $x_B \geq 0$, then the basic solution is called a basic feasible solution. If one or more variables in the basic feasible solution have a zero value, then it is called a degenerate basic feasible solution. Otherwise, it is called a non-degenerate basic feasible solution.

(ii) The maximum number of basic solutions in m linear equations "in which n is unknown"? (where $m < n$) is $\binom{n}{m}_{nC_m}$. To get all these basic solutions, every set of m columns must be linearly independent.

Example 1.7. *Find all the possible basic solutions of the following simultaneous linear equations:*

$$2x_1 + 4x_2 + x_3 = 21$$

$$x_1 + 2x_2 + 5x_3 = 16$$

Solution: The matrix representation of the given system of equations is

$$\begin{pmatrix} 2 & 4 & 1 \\ 1 & 2 & 5 \end{pmatrix} \begin{pmatrix} x_1 \\ x_2 \\ x_3 \end{pmatrix} = \begin{pmatrix} 21 \\ 16 \end{pmatrix}$$

Here the rank of the coefficient matrix A is 2. The following are our 2×2 non-singular sub-matrices from the coefficient matrix.

$$\begin{pmatrix} 2 & 1 \\ 1 & 5 \end{pmatrix} \text{ and } \begin{pmatrix} 4 & 1 \\ 2 & 5 \end{pmatrix}.$$

The sub-matrix $\begin{pmatrix} 2 & 4 \\ 1 & 2 \end{pmatrix}$ will not be considered, as it is a singular matrix.

When $B = \begin{pmatrix} 2 & 1 \\ 1 & 5 \end{pmatrix}$, we have

$$\begin{pmatrix} 2 & 1 \\ 1 & 5 \end{pmatrix} \begin{pmatrix} x_1 \\ x_3 \end{pmatrix} = \begin{pmatrix} 21 \\ 16 \end{pmatrix}$$

$$\Rightarrow \begin{pmatrix} x_1 \\ x_3 \end{pmatrix} = \begin{pmatrix} 0.56 & -0.11 \\ -0.11 & 0.22 \end{pmatrix} \begin{pmatrix} 21 \\ 16 \end{pmatrix}$$

(B^{-1} is obtained using **MINVERSE** in Excel)

$$\Rightarrow \begin{pmatrix} x_1 \\ x_3 \end{pmatrix} = \begin{pmatrix} 9.89 \\ 1.22 \end{pmatrix}$$

Hence, $x_B = (x_1 = 9.89, x_3 = 1.22)$ and $x_{NB} = (x_2 = 0)$, which is a non-degenerate basic feasible solution.

Similarly, when $B = \begin{pmatrix} 4 & 1 \\ 2 & 5 \end{pmatrix}$ bh, $x_B = (x_2 = 4.94, x_3 = 1.22)$ and $x_{NB} = (x_1 = 0)$, which is also a non-degenerate basic feasible solution.

1.4. The Gauss-Jordan Method for Solving Systems of Linear Equations

Here we shall discuss a very efficient method (the Gauss-Jordan method) for solving a system of linear equations. **Gauss-Jordan elimination** involves creating an augmented matrix of both sides of our equations, changing this matrix into reduced row echelon form (a form in which a matrix has zeros on the lower diagonal and the first non-zero number in each row is 1. Also, if a column has a leading 1, then all the other numbers in that column below 1 need to be 0), then finishing up the problem to find our solution. This method can lead us to one of the following three cases:

(i) The system has no solution.

(ii) The system has a unique solution.

(iii) The system has an infinite number of solutions.

The elementary row operation that we apply in this method is important in the sense that a similar type of elimination method will be used in the simplex method for solving a given linear programming problem (LPP).

Example 1.8. *(Problem with no solution).*

$$2x_1 + 3x_2 = 10$$
$$10x_1 + 15x_2 = 50$$

The augmented matrix representation of the previous system is:

$$A \backslash b = \begin{pmatrix} 2 & 3 & | & 10 \\ 10 & 15 & | & 60 \end{pmatrix}$$

$$= \begin{pmatrix} 1 & 1.5 & | & 5 \\ 0 & 0 & | & 10 \end{pmatrix}$$

(Divide R_1 by 2 and Multiply new R_1 by 10 and subtract from R_2)

It can be easily seen that matrix A cannot be converted to an identity matrix. This implies:

$$x_1 + 1.5x_2 = 5$$
$$0x_1 + 0x_2 = 10$$

Whatever the values of x_1 and x_2 are, the second equation can never be satisfied. Hence, the given system of equations has no solution.

Example 1.9. *(Problem with a unique solution). Use the Gauss-Jordan method to solve the following system of simultaneous linear equations:*

$$2x_1 + x_2 + x_3 = 6$$
$$-x_1 + x_2 + x_3 = 4$$
$$2x_1 + 3x_2 + x_3 = -6$$

The augmented matrix representation of the previous system is:

$$A \mid b = \begin{pmatrix} 2 & 1 & 1 & 6 \\ -1 & 1 & 1 & 4 \\ 2 & 3 & 1 & -6 \end{pmatrix}$$

$$= \begin{pmatrix} 1 & 0.5 & 0.5 & 3 \\ 0 & 1.5 & 1.5 & 7 \\ 0 & 2 & 0 & -12 \end{pmatrix}$$

(Divide R_1 by 2)

(Multiply new R_1 by 1 and add it with R_2)

(Multiply new R_1 by 2 and subtract it from R_3)

$$= \begin{pmatrix} 1 & 0 & 0 & 0.6667 \\ 0 & 1 & 1 & 4.6667 \\ 0 & 0 & -2 & -21.333 \end{pmatrix}$$

(Multiply new R_2 by 0.5 and subtract it from R_1)

(Divide R_2 by 1.5)

(Multiply new R_2 by 2 and subtract it from R_3)

(Subtract new R_3 from R_2)

(Divide R_3 by -2)

$$= \begin{pmatrix} 1 & 0 & 0 & 0.6667 \\ 0 & 1 & 0 & -6 \\ 0 & 0 & 1 & 10.667 \end{pmatrix}$$

Hence, the solution to the given system of equations is $x_1 = 0.6667$, $x_2 = -6$, and $x_3 = 10.667$. It can also be seen that the previous system of equations has a unique solution.

Example 1.10. *(Problem with an infinite solution). Use the Gauss-Jordan method to solve the following system of simultaneous linear equations:*

$$x_1 + 2x_2 + x_3 = 8$$
$$x_1 + x_2 + x_3 = 6$$
$$x_1 + x_3 = 4$$

The augmented matrix representation of the previous system is:

$$A \mid b = \begin{pmatrix} 1 & 2 & 1 & 8 \\ 1 & 1 & 1 & 6 \\ 1 & 0 & 1 & 4 \end{pmatrix}$$

$$= \begin{pmatrix} 1 & 2 & 1 & 8 \\ 0 & 1 & 0 & 2 \\ 0 & -2 & 0 & -4 \end{pmatrix} \qquad \begin{matrix} \text{(Subtract } R_2 \text{ from } R_1\text{)} \\ \text{(Subtract } R_1 \text{ from } R_3\text{)} \end{matrix}$$

$$= \begin{pmatrix} 1 & 0 & 1 & 4 \\ 0 & 1 & 0 & 2 \\ 0 & -2 & 0 & -4 \end{pmatrix} \qquad \text{(Add } R_1 \text{ and } R_3\text{)}$$

$$= \begin{pmatrix} 1 & 0 & 1 & 4 \\ 0 & 1 & 0 & 2 \\ 0 & 0 & 0 & 0 \end{pmatrix} \qquad \text{(Add } R_3 \text{ and } 2 * R_2\text{)}$$

The linear system corresponding to $A \mid b$ is

$$x_1 + x_3 = 4$$
$$x_2 = 2$$

This implies that $x_2 = 2$ and $x_1 = 4 - x_3$; that is, for different values of x_3, we can obtain different values of x_1. Hence, we have **infinite solutions** to the given system of equations.

Remark: In the Gauss-Jordan methods the following points can be noted:

(i) In the final augmented matrix, if we have any row with $[0, 0, 0, 0 \mid k]$ with $k \neq 0$, then the system of equations will not have a solution.

(ii) If the final augmented matrix is in the form $[I_n \mid k]$, then the system of equations will have a unique solution.

(iii) If we have any row with $[0, 0, 0, \ldots, 0 \mid k]$ with $k = 0$, then the system of equations will have an infinite number of solutions.

Example 1.11. *Use **Gauss-Jordan elementary row operations** to find the inverse of the matrix given in Example 1.4.*

Solution: To find inverse of A using the Gauss-Jordan method, form the augmented matrix $(A \mid I)$. **Now** we will try to reduce the matrix A to an identity matrix by elementary row operations:

$$\left(\begin{array}{ccc|ccc} 2 & 4 & 6 & 1 & 0 & 0 \\ 1 & 8 & 3 & 0 & 1 & 0 \\ 1 & 3 & 1 & 0 & 0 & 1 \end{array}\right)$$

Divide the first row by 2 and subtract the second and third rows from the new row.

$$\left(\begin{array}{ccc|ccc} 1 & 2 & 3 & 0.5 & 0 & 0 \\ 0 & 6 & 0 & -0.5 & 1 & 0 \\ 0 & 1 & 2 & -0.5 & 0 & 1 \end{array}\right)$$

Divide the second row by 6 and subtract the new row from the third row, and also multiply the new row by 2 and subtract from the first row.

$$\left(\begin{array}{ccc|ccc} 1 & 0 & 3 & 0.67 & -0.33 & 0 \\ 0 & 1 & 0 & -0.08 & 0.17 & 0 \\ 0 & 0 & -2 & -0.42 & -0.17 & 1 \end{array}\right)$$

Divide the third row by -2, multiply the new row by 3, and subtract it from the first row.

$$\left(\begin{array}{ccc|ccc} 1 & 0 & 0 & 0.04 & -0.58 & 1.5 \\ 0 & 1 & 0 & -0.08 & 0.17 & 0 \\ 0 & 0 & 1 & 0.21 & 0.08 & -0.5 \end{array}\right)$$

So the inverse of the matrix A is $\begin{pmatrix} 0.04 & -0.58 & 1.5 \\ -0.08 & 0.17 & 0 \\ 0.21 & 0.08 & -0.5 \end{pmatrix}$

1.5. Convex Sets

Line Segment: The line segment joining any two points x and y from R^n is a collection of points u, where

$$u = \lambda x + (1 - \lambda)y, \text{ for all } 0 \le \lambda \le 1.$$

Here the points x and y are called the endpoints of the line segment. It is usually denoted by $[x : y]$.

The open line segment joining x and y is a collection of points, u, where

$$u = \lambda x + (1 - \lambda)y, \text{ for all } 0 < \lambda < 1.$$

It is usually denoted by $(x : y)$.

Convex Sets: A set S is said to be a convex set, if for any two points belonging to the set, the line segment joining these two points also belongs to the set itself.

For example, for any two points x_1 and x_2 in S, the line segment joining these two points $\lambda x_1 + (1 - \lambda)x_2 \, \epsilon \, S$ for each $\lambda \epsilon \, [0, 1]$.

The line segment $\lambda x_1 + (1 - \lambda)x_2$ for $\lambda \epsilon \, [0, 1]$ is also called a convex combination of x_1 and x_2.

Fig. 1.10 Convex sets

Fig. 1.11 Non-Convex sets

Example 1.12. *Prove that* $S = \{(x_1, x_2): x_1 x_2 \geq 1; x_1 \geq 0, x_2 \geq 0\}$ *is convex.*

Solution: Let $u = (u_1, u_2)$ and $v = (v_1, v_2)$ be any two points of the set S, then

$$u_1 u_2 \geq 1 \qquad \qquad \ldots(1.2)$$

$$v_1 v_2 \geq 1 \qquad \qquad \ldots(1.3)$$

Multiplying (1.2) and (1.3),

$$u_1 u_2 v_1 v_2 \geq 1 \qquad \qquad \ldots(1.4)$$

Let $r = (r_1, r_2)$ represent a point on the line segment joining u and v. Then

$$r_1 = \lambda u_1 + (1 - \lambda) v_1, \text{ for all } 0 \leq \lambda \leq 1 \qquad \qquad \ldots(1.5)$$

$$r_2 = \lambda u_2 + (1 - \lambda) v_2, \text{ for all } 0 \leq \lambda \leq 1 \qquad \qquad \ldots(1.6)$$

Now, $r_1 r_2 = \left(\lambda u_1 + (1-\lambda) v_1\right)\left(\lambda u_2 + (1-\lambda) v_2\right)$

$$= \lambda^2 u_1 u_2 + \lambda(1-\lambda) u_1 v_2 + \lambda(1-\lambda) v_1 u_2 + (1-\lambda)^2 v_1 v_2$$

$$= \lambda^2 u_1 u_2 + \lambda(1-\lambda)[u_1 v_2 + v_1 u_2] + (1-\lambda)^2 v_1 v_2$$

$$= \lambda^2 u_1 u_2 + \lambda(1-\lambda)[(\sqrt{u_1 v_2} - \sqrt{v_1 u_2})^2 + 2\sqrt{u_1 v_1 u_2 v_2}] + (1-\lambda)^2 v_1 v_2$$

$$= \lambda^2 u_1 u_2 + 2\lambda(1-\lambda)\sqrt{u_1 v_1 u_2 v_2} + (1-\lambda)^2 v_1 v_2$$

$$\qquad \qquad + \lambda(1-\lambda)\left[(\sqrt{u_1 v_2} - \sqrt{v_1 u_2})^2\right]$$

$$\geq \lambda^2 u_1 u_2 + 2\lambda(1-\lambda)\sqrt{u_1 v_1 u_2 v_2} + (1-\lambda)^2 v_1 v_2$$

$$\left(\because \lambda(1-\lambda)\left[(\sqrt{u_1 v_2} - \sqrt{v_1 u_2})^2\right] \geq 0\right.$$

$$\geq \lambda^2 u_1 u_2 + 2\lambda(1-\lambda)\sqrt{u_1 v_1 u_2 v_2} + (1-\lambda)^2 v_1 v_2$$

$$\geq \lambda^2 + 2\lambda(1-\lambda) + (1-\lambda)^2$$

$$= 1$$

$$\Rightarrow \quad r_1 + r_2 \geq 1$$

Hence the set S is a convex set.

Example 1.13. *Show that the set* $S = \{(x_1, x_2): x^2_1 + x^2_2 \leq 9; x_1 \geq 0, x_2 \geq 0\}$ *is convex.*

Solution: Let $u = (u_1, u_2)$ and $v = (v_1, v_2)$ be any two points of the set S, then

$$u_1{}^2 + u_2{}^2 \leq 9 \qquad \qquad \ldots(1.7)$$

$$v_1{}^2 + v_2{}^2 \leq 9 \qquad \qquad \ldots(1.8)$$

Let $r = (r_1 r_2)$ represent a point on the line segment joining u and v. Then

$$r_1 = \lambda u_1 + (1 - \lambda)\, v_1, \text{ for all } 0 \le \lambda \le 1 \qquad \qquad ...(1.9)$$

$$r_2 = \lambda u_2 + (1 - \lambda)\, v_2, \text{ for all } 0 \le \lambda \le 1 \qquad \qquad ...(1.10)$$

Now, $\quad r_1^2 + r_2^2 = (\lambda u_1 + (1- \lambda)\, v_1)^2 + (\lambda u_2 + (1 - \lambda)\, v_2)^2$

$$= \lambda^2 (u_1^2 + u_2^2) + (1 - \lambda)^2 (v_1^2 + v_2^2) + 2\lambda (1 - \lambda)(u_1 v_1 + u_2 v_2)$$

$$\le 9\lambda^2 + 9 (1 - \lambda)^2 + 2\lambda (1 - \lambda)(u_1 v_1 + u_2 v_2) \qquad ...(1.11)$$

$$\text{(Using (1.7) and (1.8))}$$

Also since $u_1 v_1 + u_2 v_2 \le (\sqrt{u_1^2 + u_2^2})(\sqrt{v_1^2 + v_2^2})$ we have

$$r_1^2 + r_2^2 \le 9\lambda^2 + 9 (1 - \lambda)^2 + 2\lambda (1 - \lambda)\, 9$$

$$= 9$$

$$\Rightarrow \qquad r_1^2 + r_2^2 \le 9$$

Hence, the set S is a convex set.

Example 1.14. *Show that a line segment [x : y] joining any two points x, y $\in$ R^n is a convex set.*

Proof. The line segment joining the two points $x, y \in R^n$ is given by,

$$\{\lambda x + (1 - \lambda)\, y, \text{ where } 0 \le \lambda \le 1\} \qquad \qquad ...(1.12)$$

Let $\quad u, v \in [x : y]$, then

$$u = \lambda' x + (1 - \lambda')\, y,\, 0 \le \lambda' \le 1 \qquad \qquad ...(1.13)$$

$$v = \lambda'' x + (1 - \lambda'')\, y,\, 0 \le \lambda'' \le 1 \qquad \qquad ...(1.14)$$

Also let w denote a point on the line segment joining the two points u and v, then

$$w = \beta u + (1 - \beta)\, v,\, 0 \le \beta \le 1 \qquad \qquad ...(1.15)$$

From (1.13), (1.14), and (1.15), we have

$$w = \beta (\lambda' x + (1 - \lambda')\, y) + (1 - \beta)(\lambda'' x (1 - \lambda'')\, y),$$

$$= (\beta\lambda' + (1 - \beta)\, \lambda'')\, x + (\beta (1 - \lambda') + (1 - \beta)(1 - \lambda'')\, y) \quad ...(1.16)$$

Putting,

$$\alpha = \beta\lambda' + (1 - \beta)\, \lambda'',$$

since $0 \le \beta,\, \lambda'$ and $\lambda'' \le 1$, we have $0 \le \alpha \le 1$.

Now, $1 - \alpha = 1 - [\beta\lambda' + (1 - \beta)\,\lambda'']$

$$= \beta + (1 - \beta) - \beta\lambda' - (1 - \beta)\,\lambda''$$

$$= (\beta\,(1 - \lambda') + (1 - \beta)\,(1 - \lambda''))$$

Therefore, (1.16) can be rewritten as:

$$w = \alpha x + (1 - \alpha)\,y,\ 0 \le \alpha \le 1.$$

$\Rightarrow w \in [x : y]$, hence the line segment is a convex set.

Hyperplane: A set $S = \{x\colon c'x = \alpha\}$ is said to be a **hyperplane** in a n-dimensional space if c is a non-zero vector in R^n and α is any scalar.

A straight line in a 2-dimensional space and a plane in 3-dimensional space are examples of hyperplanes.

Theorem 1.2. A hyperplane $S = \{x\colon c'x = \alpha\}$ is a convex set.

Proof. Let $u, v \in S$, then $c'u = \alpha$

$$c'v = \alpha$$

Let w denote a point on the line segment joining the two points u and v, then

$$w = \lambda u + (1 - \lambda)\,v,\ 0 \le \lambda \le 1 \qquad\qquad ...(1.17)$$

Now, $c'w = c'\,[\lambda u + (1 - \lambda)\,v] = \lambda\,c'u + (1 - \lambda)\,c'v = \lambda\alpha + (1 - \lambda)\,\alpha = \alpha$, which implies that all the points on the line segment joining the two points u and v are also part of the set S hyperplane. Hence, a hyperplane is a convex set.

Half-Space: A closed half-space is defined by $S = \{x\colon ax \le \alpha\}$ or $S = \{x\colon ax \ge \alpha\}$, where a is any non-zero vector in Rn and α is any scalar. 2- All the points on one side of the straight line in 2-dimensional space and all the points on one side of plane in 3-dimensional space are examples of half-space.

The sets $S = \{x\colon ax < \alpha\}$ and $S = \{x\colon ax > \alpha\}$, are called open half-spaces.

Theorem 1.3. A half-space is a convex set.

Proof. Consider the closed half-space $S = \{x\colon ax \le \alpha\}$, where a is any non-zero vector in R^n and α is any scalar.

Let $u, v \in S$, such that

$$au \le \alpha \text{ and } av \le \alpha \qquad\qquad ...(1.18)$$

Let w be a point on the line segment joining the two points u and v, then

$$w = \lambda u + (1 - \lambda)\,v,\ \textit{for all } 0 \le \lambda \le 1$$

Now, $aw = a\,[\lambda u + (1-\lambda)\,v] = \lambda au + (1-\lambda)\,av \le \lambda\,\alpha + (1-\lambda)\,\alpha = \alpha$ (using (1.18))

Therefore, $aw = \le \alpha \Rightarrow$ All the points on the line segment joining the two points u and v are also part of the closed half-space. Hence, the given closed half-space is a convex set. Similarly, it can be proved for open half-spaces.

Theorem 1.4. The intersection of any two convex sets is also a convex set.

Proof: Let S_1 and S_2 be any two convex sets. And let $T = S_1 \cap S_2$.

If $x,y \in T$, then $x,y \in S_1$ and S_2. Let x' represent a point on the line segment joining x and y, then

$$x' = \lambda x + (1-\lambda)\,y, \text{ for all } 0 \le \lambda \le 1$$

Since S_1 and S_2 are convex sets, $x' \in S_1$ and S_2, which implies $x' \in T$ **also.**

$\Rightarrow T = S_1 \cap S_1$ is a convex set.

Remark: *The intersection of any finite number of convex sets is again a convex set.*

Polyhedron. It is the intersection of a finite number of half-spaces.

Theorem 1.5. The sum and difference of any two convex sets is again a convex set.

Proof. Let A and B be any two convex sets in $\mathbf{R}^n$, and then we have to show that $A \pm B$ is also a convex set.

Let u and v be any two points of the sets $A \pm B$ so that

$u = x_1 \pm y_1$ and $v = x_2 \pm y_2$, where $x_1, x_2, \in A$ and $y_1, y_2 \in B$. Let w be any point on the line segment joining u and v, then

$$w = \lambda u + (1-\lambda)\,v, \text{ for all } 0 \le \lambda \le 1$$
$$= \lambda\,(x_1 \pm y_1) + (1-\lambda)\,(x_2 \pm y_2)$$
$$= [\lambda x_1 + (1-\lambda)\,x_2] \pm [\lambda y_1 + (1-\lambda)\,y_2]$$

"Since A is a convex set and $x_1, x_2 \in A$, we have"?

$$\lambda x_1 + (1-\lambda)\,x_2 \in A, \text{ for all } 0 \le \lambda \le 1$$

Similarly, for $y_1, y_2 \in B$,

$$\lambda y_1 + (1-\lambda)\,y_2 \in B, \text{ for all } 0 \le \lambda \le 1$$

This implies,

$$w = \lambda u + (1-\lambda)\,v \in A \pm B, \text{ for all } 0 \le \lambda \le 1.$$

Thus, for any two points u and v from the set $A \pm B$, the line segment joining these two points is also in $A \pm$ B. Hence, the sum and difference of any two convex sets is again a convex set.

Convex Hull. A convex hull of a set C of "n" points from R^n, denoted by $H(C)$, is the smallest perimeter fence in R^n enclosing these "n" points.

If C is a convex set, then $H(C) = C$. The following are some illustrations of convex hulls.

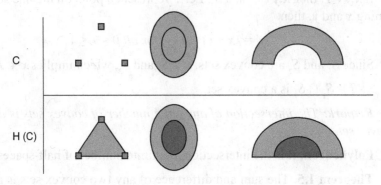

Fig. 1.12

Hence, a convex hull can also be defined as:

- the smallest convex set containing all the points
- the smallest area convex "polygon" enclosing the points
- a convex "polygon" enclosing the points, whose vertices are points in the set

Convex combination of vectors. Let $S = \{x_1, x_2, x_3, ..., x_m\}$ be a set of "m" vectors from R^n. Then a linear combination of these vectors, $x = \sum_{i=1}^{m} \lambda_i x_i$, where $\lambda_i \geq 0$ and $\sum_{i=1}^{m} \lambda_i = 1$, is called a convex combination of the given vectors.

Convex polyhedron. The set of all the convex combination of a finite number of vectors in R^n is called a convex polyhedron or a polytope spanned by these vectors. In other words, a polytope is a bounded polyhedron and always forms a convex set.

Simplex. A simplex in k-dimension is a polytope having exactly $(k + 1)$ vertices. A simplex in a 1-dimensional space is a line segment, in two dimensnsions it is a triangle, and so on.

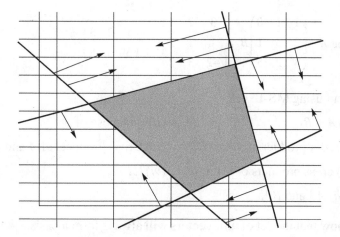

Fig. 1.13 Polytope (a bounded Polyhedron)

Extreme Point: Let S be a convex set. A point $r \in S$ is called an extreme point if, for any two points $u, v \in S$ (where $u \neq v$), r cannot be written as a convex combination of the points u and v. An extreme point will always be a boundary point, but all boundary points will not be extreme points.

 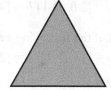

Fig. 1.14

For example, a circle is a convex set, and all the points on the circumference of the circle are boundary as well as extreme points of the set, whereas in the case of a triangle, the three corner points are the extreme points of the set.

Exercises

1. Define (*i*) basic solution, (*ii*) convex set, (*iii*) convex polyhedron, and (*iv*) extreme points of a convex set.

2. Define a line segment and show that a line segment is a convex set.

3. Let $A = \begin{pmatrix} 1 & 1 & 0 \\ 2 & 3 & 1 \\ 3 & 2 & 5 \end{pmatrix} B = \begin{pmatrix} 3 & 2 & 1 \\ 1 & 1 & 1 \\ -2 & 3 & 2 \end{pmatrix} C = \begin{pmatrix} 1 & 2 & 3 \\ 3 & 6 & 8 \end{pmatrix}$

Find using MS-Excel:

(*i*) $A + B$ (*ii*) transpose of A

(*iii*) dot product of A and B (*iv*) Cross Products of A and B

(*v*) cross products of C and B (*vi*) $|A|$

(*vii*) A^{-1} and B^{-1}.

4. Show that the set of unit vectors will always form a basis.

5. Show that any two bases in R^n will contain the same number of vectors.

6. Check whether the following vectors are linearly independent:

(*i*) $a = [1\ 2\ 2]$, $b = [3\ 4\ 4]$, $c = [1\ 0\ 0]$

(*ii*) $a = [2\ 1]$, $b = [4\ 3]$

(*iii*) $a = [3\ 1\ 2]$ $b = [2\ 0\ 0]$ $c = [-1\ -1\ -2]$

(*iv*) $a = [2\ 0\ -3\ 1]$ $b = [4\ 0\ -6\ 2]$

7. Check if the following vectors form the basis:

(*i*) $a = [1\ -1\ -1]$, $b = [2\ -1\ 3]$, $c = [1\ 2\ -1]$

(*ii*) $a = [1\ 2\ 3]$, $b = [1\ -1\ -1]$ $c = [1\ 2\ -1]$

(*iii*) $a = [4\ 3\ 2\ -1]$ $b = [4\ 2\ 0\ -2]$

8. Check with the help of the rank method whether the following system of equations are consistent and solve them by using the Gauss-Jordan method:

(*i*) $0.5x_1 + 0.5x_2 + 0.5x_3 = 2$, $4x_1 + 10x_2 - 4x_3 = 6$, $x_1 + 7x_2 - 7x_3 = 5$

(*ii*) $x_1 - x_2 + x_3 = 2$, $-1.5x_1 + 0.5x_2 - x_3 = 3$, $-3x_1 - x_2 - x_3 = 18$

(*iii*) $x_1 - x_2 + x_3 = 7$, $x_1 - 2x_2 + 3x_3 = 8$, $3x_1 + 6x_2 + 9x_3 = 24$

9. Find the inverse of the matrices A, B, C given in Question 3 by the Gauss-Jordan Method.

10. Determine how many basic solutions exist for the following system of equations and find all basic feasible solutions:

   (*i*) $x_1 + 2x_2 + 10x_3 + 4x_4 = 8, x_1 + x_2 + 4x_3 + 3x_4 = 11$

   (*ii*) $x_1 + x_2 + x_3 = 7, x_1 + 2x_2 + 3x_1 = 8,$

   (*iii*) $4x_1 + 10 x_2 - 4x_3 = 6, x_1 + 7x_2 - 7x_3 = 5,$

11. Check if the following sets are convex:

   (*i*) $S = \{[x_1, x_2], x^2_1 + 3x^2_2 \le 8\}$

   (*ii*) $S = \{[x_1, x_2], x_1 \ge 5, x_2 \le 9\}$

   (*iii*) $S = \{[x_1, x_2], \left(x_1 - \dfrac{3}{2}\right)^2 + (x_2 + 5)^2 \le 10\}$

   (*iv*) $S = \{[x_1, x_2], -x_1 + x_2 \le 2, 2x_1 + 3 x_2 < 11, x_1 \ge 0, x_2 \ge 0\}$

   (*v*) $S = \{x_1 x_2 \le 1; x_1 \ge 0, x_2 \ge 0\}$

12. Show that the intersection of a finite number of convex sets is also a convex set.

13. Show that a convex polyhedron always forms a convex set.

2

INTRODUCTION TO LPPs AND THE GRAPHICAL METHOD

2.1. Introduction

A programming problem is a problem where we have an objective which is to be fulfilled under certain restrictions. The moment this objective and these restrictions are converted into a mathematical expression and mathematical equations or inequalities, respectively, the problem is called a mathematical programming problem. When this mathematical expression and these equations or inequalities are linear in form, that is, the relationship between the variables is linear, in which each variable will only have a power of one and product of variables, logarithmic, exponential, trigonometric, etc. terms are not allowed, then the mathematical programming problem is called a linear programming problem (LPP). Hence, a linear programming problem is a special type of mathematical programming problem which either maximizes or minimizes a linear objective function while satisfying a set of linear equality and/or inequality constraints (restrictions). Mathematical programming problems are optimization problems. In general, a problem where we maximize or minimize a numerical function of variables under some constraints or restrictions which are also in the form of the numerical functions is called an optimization problem. These types of problems may have a number of feasible solutions which satisfy all the constraints, out of which one is selected, which optimizes our objective function.

Any general linear programming problem with n variables and m linear constraints can be written as:

Maximize or Minimize $Z = c_1 x_1 + c_2 x_2 + \dots + c_n x_n$ (Linear objective function)

Subject to the constraints

$$a_{11}x_1 + a_{12}x_2 + \ldots + a_{1n}x_n \left(\leq, =, \geq\right) b_1$$
$$a_{21}x_1 + a_{22}x_2 + \ldots + a_{2n}x_n \left(\leq, =, \geq\right) b_2$$
$$\vdots \qquad \vdots \qquad \vdots$$
$$a_{i1}x_1 + a_{i2}x_2 + \ldots + a_{in}x_n \left(\leq, =\geq\right) b_i$$
$$\vdots \qquad \vdots \qquad \vdots$$
$$a_{m1}x_1 + a_{m2}x_2 + \ldots + a_{mn}x_n \left(\leq, =\geq\right) b_m$$

...(2.1)

$$x_1, x_2, \ldots, x_n \geq 0 \text{ (Non-negative restriction)}$$

The linear inequalities and/or equations represent the restrictions on available resources like man-hours, machine-hours, land, and so on Here, a_{ij}, b_i, c_j where $i = 1, 2....m$, $j = 1, 2...n$ are assumed to be constants. The coefficients c_1, c_2, ..., c_n are the unit profits/ cost coefficients of the decision variables x_1; x_2, ..., x_n, respectively, in the objective function. The inequality $\sum_{j=1}^{n} a_{ij}x_j \leq b_i$ ($i = 1, 2, \ldots, m$) represents the ith functional constraint (restriction) on the consumption of the ith resource where a_{ij} the coefficient of x_j ($j = 1, 2, \ldots, n$) gives the amount of resource i consumed by one unit of activity j, x_j is the level of activity j, and b_i is the maximum availability of the ith resource.

The objective is to determine the non-negative values of the variables x_1; x_2, ..., x_n which will satisfy all the constraints on available resources and optimizes the objective function.

(2.1) can also be written as:

Maximize or Minimize $Z = \sum_{j=1}^{n} c_j x_j$

Subject to $\sum_{j=1}^{n} a_{ij} x_j (\leq, =, \geq) b_i$, $i = 1, 2 \ldots m$ ...(2.2)

$$x_j \geq 0, j = 1, 2 \ldots n$$

Or in matrix form, the same can be written as:

Maximize or Minimize $Z = \mathbf{CX}$

Subject to $AX (\leq, =, \geq) \mathbf{b}$

$$X \geq 0$$

...(2.3)

Here, $b_{m \times 1} = (b_1, b_2 \ldots\ldots b_m)$, $C_{(1 \times n)} = (c_1, c_2, \ldots c_n)$, $X'_{(1 \times n)} = (x_1, x_2 \ldots x_n)$

$$A_{(m \times n)} = \begin{pmatrix} a_{11} & \cdots & a_{1n} \\ \vdots & \ddots & \vdots \\ a_{m1} & \cdots & a_{mn} \end{pmatrix}$$

Solution: An *n-tuple* $(x_1, x_2, \ldots, x_n)$ of real numbers which satisfies the constraints of (2.1) is called a solution to the general linear programming problem.

Feasible Solution: A solution to a linear programming problem is known as a feasible solution if it also satisfies the non-negative restriction of the problem.

Basic Feasible Solution: If a basic solution to the given linear programming problem satisfies the non-negativity conditions also then the solution is called a basic feasible solution. Basic feasible solutions are of two types:

(i) Degenerate: A basic feasible solution is called degenerate if the value of at least one basic variable is zero.

(ii) Non-degenerate: A basic feasible solution is called "non-degenerate" if values of all the variables are greater than zero.

Optimal Basic Feasible solution: If a basic feasible solution to a given linear programming problem optimizes the objective function, then it is called an optimal basic feasible solution.

Note. If a problem has a feasible solution, then it will also have a basic feasible solution.

Example: *Minimize* $\quad z = x_1 + 2x_2$

Subject to

$$x_1 + 2x_2 \geq 10$$
$$10x_1 + 3x_2 \geq 20$$
$$x_1, x_2 \geq 0$$

This is a minimization problem with two decision variables and two constraints. Here, $c_1 = 1$, $c_2 = 2$, $b_1 = 10$, $b_2 = 20$, $a_{11} = 1$, $a_{12} = 2$, $a_{21} = 10$, and $a_{22} = 3$.

$x_1 = -1$, $x_2 = 15$ is a solution to the given problem, but it is not a feasible one, as $x_1 < 0$.

$x_1 = 0.588235$, $x_2 = 4.705882$ is an optimal basic feasible solution to the given problem, and the corresponding minimum value of z is 10.

2.2. Assumptions in a Linear Programming Problem

To solve any real-life problem by the linear programming approach (LP), the following assumptions are made:

1. Certainty. This assumption requires that all the parameters of the model be known before solving a given LP problem. In other words, the coefficients c_j of the objective function, the coefficients of the constraints a_{ij}, and the resource vector b_i must all be known and fixed"? must all be known and fixed. However, using sensitivity analysis (which will be discussed in Chapter 6), the decision maker can explore the effects of changing some parameters over a range of values.

2. Linearity. This is a property of the mathematical function. Linearity denotes a stable relationship between a dependent and independent variables, which when plotted on a graph would be represented by a straight line. In linear programming, the relationship of the various decision variables either in the objective function or in any of the constraints should be linear. Note that in a linear relationship, all variables have an exponent of one, and no variables are put in product form.

The following are three properties which help us in clarifying the implications of linearity:

(i) **Proportionality.** This requires that the value of the objective function be proportional to the value of the decision variables. For example, in an objective function *Max* $25x_1 + 38x_2$, each unit of x_1 should contribute 25 (a proportionality constant) to the value of the objective function, and hence the total contribution of any activity in the objective function depends on the level of activity or the value of the decision variable. In the constraints, the amount of each resource used must be proportional to the value of the decision variable. For example, let $x_1 + 4x_2 \leq 50$ be any constraint of a given LPP, and then each value of x_2 will consume 4 units (a proportional amount) of the available resource (in this case 50).

(ii) **Additivity.** Linearity also requires that the effect of the value of each variable on the values of the objective function and the constraints is additive. In other words, there can be no interactions between the effects of different activities. For some objective function, for example, *Max* $25x_1 + 38x_2$, if $x_1 = 2$ and $x_2 = 1$, this property asserts that the contributions of 50 and 38, respectively, must add together to give a sum of 88.

(iii) **Divisibility (or continuity).** This postulate allows the variables to have a continuous range of values. This means that the variables may take any fractional or decimal value and will not necessarily be a whole number or integer. However, in many real-life situations, there may be cases where the variables are allowed to have only integer values, and in such situations LP may be used to provide an approximate answer by rounding off the solution to the nearest integer value, but doing so may cause a significant departure from optimality. If integer solutions are required, then the given linear programming problem is a special type of problem called a linear integer programming problem (to be discussed later).

2.3. Theorems on Extreme Points

Theorem 2.1. The set of feasible solutions to a given LPP forms a convex set.

Proof. Let S denote the set of feasible solutions to the following LPP,

$$Max\ Z = CX$$

Subject to

$$AX \leq b$$

$$X \geq 0 \qquad \qquad \ldots(2.4)$$

where $\qquad S = \{X \mid AX \leq b, X \geq 0\}$

Let $X_1, X_2 \in S$, such that

$$AX_1 \leq b \qquad \qquad \ldots(2.5)$$

$$AX_2 \leq b \qquad \qquad \ldots(2.6)$$

Let X' denote the point on the line segment joining X_1 and X_2, then

$$X' = \lambda X_1 + (1 - \lambda)X_2,\ 0 \leq \lambda \leq 1 \qquad \qquad \ldots(2.7)$$

Multiplying both sides of (2.7) by A, we get

$$AX' = \lambda AX_1 + (1 - \lambda)AX_2 < \lambda b + (1 - \lambda)b \qquad \text{(From (2.5) and (2.6))}$$

$$\Rightarrow \qquad AX' \leq b$$

which implies that $X' \in S$. Hence, the set of feasible solutions to a given LPP is a convex set.

Theorem 2.2. If the set of feasible solutions to a given LPP is a convex polyhedron, then at least one extreme point of the problem will be the optimum solution of the LPP.

Proof. Let,

$$Max\ z = cx,$$

$$Subject\ to\ Ax = b,\ x \geq 0 \qquad \qquad ...(2.8)$$

be a given LPP and S be a set of feasible solutions to (2.8), which is a convex polyhedron. Also let $x_1, x_2, ..., x_k$ denote the extreme points of S and $x^* \in S$ be the optimum solution of (2.8), that is, $z^* = cx^* \geq cx \ \forall\ x$.

If x^* is not an extreme point of S, then

$$x^* = \sum_{i=1}^{k} \lambda_i x_i, \lambda_i \geq 0 \text{ and} \sum_{i=1}^{k} \lambda_i = 1 \qquad \qquad ...(2.9)$$

This implies $z^* = cx^* = \sum_{i=1}^{k} \lambda_i cx_i$ $\qquad \qquad ...(2.10)$

Let max $cx_i = cx_p \ \forall\ i = 1, 2, ..., k$. Now (2.10) can be written as

$$cx^* \leq \left(\sum_{i=1}^{k} \lambda_i\right) cx_p \Rightarrow cx^* \leq cx_p, \text{ since } \sum_{i=1}^{k} \lambda_i = 1$$

But cx^* is the maximum value of z, hence $cx^* = cx_p$

$\Rightarrow \qquad x^* = x_p$ (one of the extreme points).

Hence, the optimum solution of the LPP is attained at one of the extreme points.

Theorem 2.3. Let $S = \{x \in R^n : Ax = b, x \geq 0\}$ be the set of feasible solutions of the following LPP,

$$Max\ z = cx,$$

Subject to

$$Ax = b, x \geq 0$$

Then $x^* \in S$ is a basic feasible solution of the given problem if and only if it is an extreme point of S.

Proof. (Part 1) Let $x^* \in S$ be a basic feasible solution of the given problem, such that

$$Ax^* = b, x^* \geq 0 \qquad \qquad ...(2.11)$$

We have to prove that this is an extreme point; that is, we have to show that two points do not exit in the set such that x^* can be written as a convex combination of these points.

On the contrary, let us assume that x^* is not an extreme point of the set of feasible solutions, and then two solutions x' and $x'' \in S$ exist, such that

$$x^* = \lambda x' + (1 - \lambda)x'',\ 0 < \lambda < 1 \qquad \text{...(2.12)}$$

Further, let $x^* = [x_B^*, x_{NB}^*], x' = [x_B', x_{NB}'], x'' = [x_B'', x_{NB}'']$ \qquad ...(2.13)

where x_B^*, x_B' and x_B'' are m component vectors and x_{NB}^*, x_{NB}', and x_{NB}'' are $(n - m)$ component vectors.

From (2.12) and (2.13), we have

$$[x_B^*, x_{NB}^*] = \lambda[x_{NB}', x_{NB}'] + (1 - \lambda)[x_B'', x_{NB}''] \qquad \text{...(2.14)}$$

Since $x^* \in S$ is a basic feasible solution, we have $x_{NB}^* = 0$ and $Bx_B^* = b$ where B is a $m \times m$ order basic matrix for the given LPP.

Now (2.14) can be written as

$$0 = \lambda x_{NB}' + (1 - \lambda)x_{NB}'' \text{ where } 0 < \lambda < 1 \Rightarrow x_{NB}' = x_B'' = 0 \quad \text{...(2.15)}$$

Further, $Ax' = Bx_B' = b$ and $Ax'' = Bx_B'' = b \qquad \Rightarrow x_B^* = x_B' = x_B''$

and hence $x^* = x' = x''$ which proves that our assumption is wrong and x^* is an extreme point of S.

(Part 2) Here we will show that if $x^* \in S$ is an extreme point, then it is a basic feasible solution to the given LPP.

Here, x^* is an n component vector. Let us assume that its r components are positive and the rest are zero such that

$$\sum_{j=1}^{r} a_j x_j^* = b,\quad x_j^* > 0,\quad j = 1, 2 \ldots r \qquad \text{...(2.16)}$$

In order to prove that x^* is a basic feasible solution, we have to show that the vectors associated with the positive components of x^* are linearly independent. To prove it, let us assume on the contrary that the vectors associated with the positive components of x^* are linearly dependent, that is:

$$\sum_{j=1}^{r} \beta_j a_j = 0 \text{ in which at least one } \beta_j \text{ is non-zero} \qquad \text{...(2.17)}$$

Now consider $\theta = min\left[\dfrac{x_j^*}{|\beta_j|}, \beta_j \neq 0, j = 1, 2, \ldots r\right]$ positive number

If we take $0 < \varepsilon < \theta$, then $x_j^* - \varepsilon\beta_j > 0$, and $x_j^* - \varepsilon\beta_j > 0, j = 1, 2 \ldots r$ ...(2.18)

In the vector form, (2.18) can be written as

$x' = x^* + \varepsilon\beta$ and $x'' = x^* + \varepsilon\beta$ where x^*, x', x'', β are n vector components. ...(2.19)

Also we have x', $x'' \geq 0$, *that is, both are feasible solutions, and hence we have*

$Ax' = Ax^* + \varepsilon\ A\beta = Ax^* = b$ (since $A\beta = 0$ from (2.17)). Similarly, $Ax'' = Ax^* + \varepsilon A\beta = Ax^* = b$

Further from (2.19) we can write $x^* = \left(\dfrac{1}{2}\right)x' + \left(\dfrac{1}{2}\right)x''$, which contradicts the fact that x^* is an extreme point. Hence, our assumption is wrong.

It proves that the vectors associated with the positive variables are linearly independent, and hence the extreme point is a basic feasible solution.

Theorem 2.4. If an optimum value of the objective function occurs at any two extreme points of the feasible region, then all the points on the line segment joining these points also gives the optimal value of the objective function.

Proof. Let the maximum value of the objective function of the following LPP,

$$Max\ z = cx,$$

Subject to
$$Ax = b, x \geq 0$$

occur at the extreme points x_1 and x_2, **and then let Max $z = z^* = cx_1 = cx_2$.** Let x' be the point on the line segment joining x_1 and x_2, and then

$$x' = \lambda x_1 + (1 - \lambda)x_2,\ 0 \leq \lambda \leq 1$$

$\Rightarrow \qquad cx' = \lambda cx_1 + (1 - \lambda)cx_2 = z^*$

$\Rightarrow \qquad x'$ is also an optimal solution of the problem and hence the theorem.

2.4. Areas of Application of LPPs

I. Product Mix Problem. Here we have to determine the right mix and quantities of products to be manufactured so as to maximize the profit. A firm can always produce different types of products which require different resources in different quantities, and each product is contributing a certain amount towards profit. The final product mix must take into consideration the limited resources, expected demand for each product, and various policies of the company, and at the same time maximize the profit.

II. Blending Problem. In this type of problem, several raw materials are mixed into the final product that must fulfill certain specifications. Each of the raw materials contributes certain properties to the final product and involves a certain cost.

III. Media Allocation Problem. Suppose a company has a certain fixed budget which is to be utilized for the advertisement of a particular product in different media where each medium has a different impact and reach. The cost of insertion of an advertisement into each medium is different and known. The problem is to determine the number of allocations of an advertisement in different media so that there is a maximum exposure of the product to its target populations, which maximizes its sales.

IV. Production Scheduling and Inventory Planning. Suppose the demand for a particular product fluctuates over a period of time. The varying production rates have proven to be costly. The problem is to determine a smooth production schedule that meets the anticipated demand and yet maintains a reasonable inventory level that minimizes the overall production and inventory carrying cost.

V. Purchasing Problem. Linear programming can also be used in deciding about the purchase of various products which are available at different discounts and prices for different quantity ranges. The problem is to determine the quantity of different products that should be purchased so as to maximize (minimize) the overall profit (cost).

VI. Assignment and Transportation Problem. The transportation and assignment problem can always be formulated as a special type of linear integer programming problem, satisfying the various row and column requirements and finally minimizing the cost of assignment or transportation.

2.5. Formulation of Linear Programming Models

The following steps need to be taken for a successful formulation of an LP model:

I. Understand the problem.

II. Identify the decision variables involved in the problem.

III. Identify the various coefficients of the objective function.

IV. Identify the various restrictions of the problem.

V. Represent the objective function in the form of a linear expression involving decision variables.

VI. Represent all the restrictions in the form of linear expressions involving decision variables.

Example 2.1. *(Product Mix) A small plant manufactures two models of a product, say A and B. Two resources, R1 and R2, are required for the manufacturing of units of this product. One unit of A consumes 10 units of R1 and 15 units of R2. Similarly, one unit of B consumes 12 units of R1 and 17 units of R2. Suppose further that the company has 800 units of R1 and 600 units of R2 available per week, and each unit of model A and B gives a profit of $50 and $70, respectively. Formulate this as a linear programming problem so as to find the optimal weekly production levels for these two models.*

Solution: Let x_1 and x_2 denote the weekly production levels of model, *A* and model *B*, respectively.

The LP model of the given problem is,

$$\textbf{\textit{Max z} = 50x_1 + 70x_2 \textbf{ (Objective function)}}$$

Subject to

$$10x_1 + 12x_2 \leq 800 \textbf{(R1 resource)}$$

$$15x_1 + 17x_2 \leq 600 \textbf{(R2 resource)}$$

$$x_1, x_2 \geq 0 \textbf{ (non-negative condition)}$$

Example 2.2. *(Production Scheduling) A firm manufactures three different versions of the same product, which have unit profit contributions of 2, 3, and 3.5. In the final part of the manufacturing process, there are assembly, polishing, and packing operations. For each version, the time (in hrs.) required for these operations is shown in the following table:*

		Assembly	Polishing	Packing
	1	4	3	3
Product versions	2	5	1	4
	3	2	4	1.5

Under the given state of the labor force, the firm estimates that each year, they have 1,000 hours of assembly time, 800 hours of polishing time, and 950 hours of packing time available. Formulate a linear programming problem to find the number of each version the company should make per year in order to maximize the associated profit.

Solution: Let x_1, x_2, and x_3 be the number of units of the first, second, and third version of the product produced per year, respectively. Then the LP model of the given problem is,

$$Max\ z = 2x_1 + 3x_2 + 3.5x_3 \text{ (Objective function)}$$

Subject to

$$4x_1 + 5x_2 + 2x_3 \leq 1000 \text{ (Assembly)}$$

$$3x_1 + x_2 + 4x_3 \leq 800 \text{ (Polishing)}$$

$$3x_1 + 4x_2 + 1.5x_3 \leq 950 \text{ (Packing)}$$

$$x_1, x_2, x_3 \geq 0 \text{ (Non-negativity condition)}$$

Example 2.3. *(Purchasing Problem) A firm is planning to buy two types of cupboards for its new office room, which has a maximum area of 110 square feet for the cupboards. The first type of cupboard costing $150 per unit requires five square feet of floor space, and holds nine cubic feet of files, whereas the second cupboard costs $170 per unit, requires seven square feet of floor space, and holds eleven cubic feet of files. The firm has a total budget of $10,000 for this purpose. How many of which type of cupboard should the firm buy in order to maximize storage volume?*

Solution: Let x_1 and x_2 be the number of units of the first and second type of cupboards, respectively. Then the objective function of the problem is to maximize storage volume of the office room subject to the cost and space constraints. So, the LP model of the given problem is,

$$Max\ z = 9x_1 + 11x_2$$

Subject to

$$150x_1 + 170x_2 \leq 10000 \text{ (Cost Restriction)}$$

$$5x_1 + 7x_2 \leq 110 \text{ (Floor Restriction)}$$

$$x_1, x_2, \geq 0 \text{ (Non-negativity condition)}$$

Example 2.4. *(Transportation Problem) A consumer durable company which manufactures a single product has three plants, I, II, and III.*

The amounts of units produced in these plants are 100, 150, and 120, respectively. The firm has its commitment to sell 80 units to customer A, 100 to B, 70 to C, and 120 to D. The net unit cost of transportation to these four different customers from the three plants is given as follows:

		Customers			
		A	B	C	D
Plants	I	10	20	15	18
	II	21	10	19	11
	III	18	14	22	13

Formulate the previous transportation problem as a linear programming problem so as to minimize the total cost of transportation.

Solution: Let x_{ij} ($i = 1, 2, 3$ and $j = 1, 2, 3, 4$) denote the amount of units transported from the ith plant to the jth customer. Then the objective function of the problem is,

$$\text{Min } z = 10x_{11} + 20x_{12} + 15x_{13} + 18x_{14} + 21x_{21} + 10x_{22}$$
$$+ 19x_{23} + 11x_{24} + 18x_{31} + 14x_{32} + 22x_{33} + 13x_{34}$$

Subject to

Row Constraints:

$$\sum_{j=1}^{4} x_{1j} = 100$$

$$\sum_{j=1}^{4} x_{2j} = 150$$

$$\sum_{j=1}^{4} x_{3j} = 120$$

Column Constraints:

$$\sum_{i=1}^{3} x_{i1} = 80$$

$$\sum_{i=1}^{3} x_{i2} = 100$$

$$\sum_{i=1}^{3} x_{i3} = 70$$

$$\sum_{i=1}^{3} x_{i4} = 120$$

and $x_{ij} \geq 0$, $\forall\ i, j$ **(non-negativity condition)**

Example 2.5. *(Media Allocation Problem) A company has an advertising budget of $10,000,000 and is planning to advertise its products in three different media, namely television, radio, and newspapers. The advertising in television is further divided into prime time and non-prime time. The objective of the company is to have maximum exposure of the product to its potential customers. A survey of the market gives the following results:*

| | Television | | Radio | News paper |
	Prime time	Non-prime time		
Cost of an Advt./unit($)	50,000	28,000	20,000	19,000
No. of potential customers reached/unit	700,000	350,000	250,000	180,000
No. of female potential customers reached/unit	500,000	100,000	110,000	87,000

It was also decided that the company will spend no more than 50% of the total budget in television. The management was also not willing to have more than 7 and 10 insertions in radio and newspapers respectively. Finally, the company has also decided to have a minimum of 1.5 million exposures among women. Formulate this problem as an LP model so as to maximize the total reach of the potential customer.

Solution: Let x_1, x_2, x_3, and x_4 denote the number of insertions in prime time, non-prime time (TV), radio, and newspapers respectively. The objective of the company is to maximize the total reach of the potential customer. Then the LP model is:

$$Max\ z = 700000x_1 + 350000x_2 + 250000x_3 + 180000x_4$$

(Maximization of potential reach)

Subject to

$$50000x_1 + 28000x_2 + 20000x_3 + 19000x_4 \leq 1,000,000,$$

(Budget constraint)

$$50000x_1 + 28000x_2 \leq 500,000,$$

(Restriction on Television advertisement)

$$500000x_1 + 100000x_2 + 110000x_3 + 87000x_4 \geq 1,500,000$$

(Restriction on the minimum number female potential customers to be reached by the advertisement)

$$x_3 \leq 7, x_4 \leq 10$$

(Upper limits on the number of insertions in Radio and Newspapers)

$$x_1, x_2, x_3, \text{ and } x_4 \geq 0$$

2.6. Graphical Method

The graphical method can be used to solve any LPP with two variables only. There are two different approaches for solving an LPP using the graphical method:

2.6.1. Extreme Point Approach

The following steps need to be followed while using this approach:

I. First, draw an *XY* plane and select the set of points (x, y) which satisfy the non-negative restrictions of the given LPP. This set of points will be a set of feasible solutions to the given problem. All these points in the set of feasible solutions lie in the first quadrant of the *xy* plane.

II. Draw all the constraints in the *xy* plane and identify the common region, that is, a set of solutions (x, y) which satisfies all the constraints and lie in the first quadrant. This set of solutions is called a common region of feasible solutions.

III. Identify the corner points (extreme points) of the common region of feasible solutions; find the value of the objective function at these points. Select the point which optimizes the objective function, and this will represent our optimal solution.

IV. Give your interpretation to the result.

Note: The common feasible region of any LP problem always forms a convex set, and its optimum solution lies at one of the extreme points of the convex set.

2.6.2. ISO-Profit (cost) Function Line Approach

After doing the first two steps (I & II) of the procedure given in 2.6.1, continue with the following:

I. Draw an ISO-profit (ISO-cost) line for a small value of the objective function without violating any of the given restrictions of the given LPP.

II. Continue drawing ISO-profit (ISO-cost) lines parallel to one another in the direction of increasing (or decreasing) the value of the objective function.

III. The last point of the feasible region where this line will touch before it moves out of the (feasible) region for which the value of ISO-profit (ISO-cost) is maximum (minimum) is the optimal solution.

We will discuss the steps involved in solving a simple linear programming model graphically with the help of an example.

In this book, our main focus will be on the extreme point approach.

2.7. Solution of LPPs by the Graphical Method Using MS-Excel

Step 1. Select the xy-values which are to be plotted and then go to **Insert → Scatter → Scatter with straight line and Markers.**

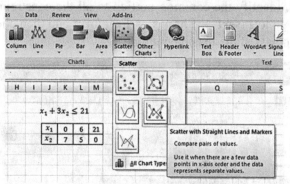

Fig. 2.1

Step 2. Right-click on the chart and click the **Select Data** option.

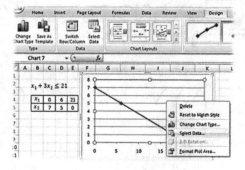

Fig. 2.2

Step 3. In the **Legend Entries (series)**, select the **Add** option and add all the constraints of the problem.

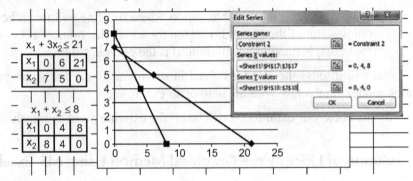

Fig. 2.3

Step 4. Identify the common portion to all the constraints. For shading this region go to **Insert → Shapes → Freeform** and then click the cursor to the corner points of the feasible region.

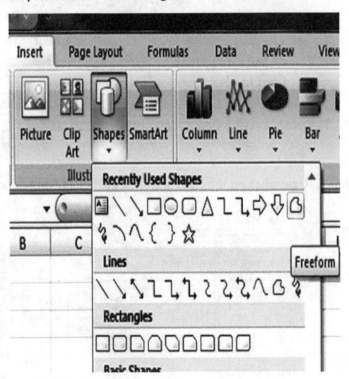

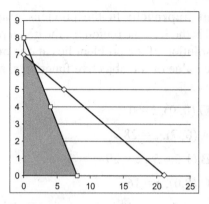

Fig. 2.4

Step 5. For the formatting of the chart, such as inserting text, gridlines, legends, and so on, double-click on the chart and select the **Layout** option.

Example 2.6. *Maximize* $\qquad z = 2x_1 + 4x_2$

Subject to the constraints:

$$x_1 + 3x_2 \leq 21$$
$$x_1 + x_2 \leq 8$$
$$x_1 \leq 3$$
$$x_1, x_2 \geq 0$$

Solution: In the xy-plane, we take x_1 on the x-axis and x_2 on the y-axis. When the two constraints along with the non-negativity conditions are plotted in the x_1x_2 plane, we have the following figure.

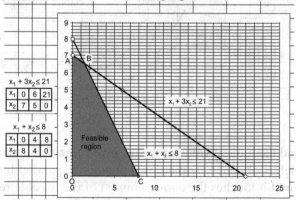

Fig. 2.5

Extreme point approach: In the previous graph the shaded region denoted by **OABC** is the feasible region for the given problem. We also know that the optimum solution of the problem lies at one of the extreme points of this region. So, the value of the objective function z. at these extreme points is:

I. at the point O $(0, 0)$, $z = 0$

II. at the point A $(0, 7)$, $z = 28$

III. at the point B $(1.5, 6.5)$, $z = 29$

IV. at the point C $(8,0)$, $z = 16$

We can clearly see that $z = 29$ is the maximum value of z, and this occurs at point B. Hence, the optimum solution of the problem is $(x_1, x_2) = (1.5, 6.5)$, and the corresponding maximum value of the objective function is $z^* = 29$.

ISO-profit (cost) function line approach: To get an **iso profit line**, set the objective function equal to some arbitrary small number. This will give us a linear equation in x and y, and that can be plotted in the $xy - plane$. All the points on an **iso profit line** give the same profit. To start with, let us put $z = 10$. Then the equation becomes $2x_1 + 4x_2 = 10$, and this line contains all the points that have a profit of exactly 10.

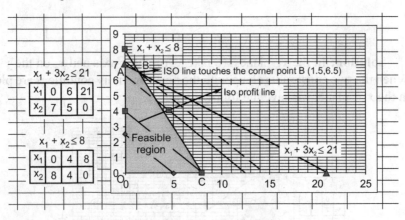

Fig. 2.6

Now find the highest value of the **iso profit line** that touches the feasible region, which is obtained by plotting a number of parallel lines, away from the origin until it touches the corner point(s) of the feasible region. Here the **iso profit line** touches the corner point **B** for the first time. The coordinates of corner points can be read from the graph or can be computed as the intersection of the two linear equations.

Therefore, the optimum solution of the problem is $x_1 = 1.5$, $x_2 = 6.5$, and the corresponding maximum value of the objective function is $z^* = 29$.

Example 2.7. *Minimize* $z = 25x_1 + 38x_2$

Subject to the constraints:

$$30x_1 + 10x_2 \geq 120$$
$$10x_1 + 20x_2 \geq 100$$
$$x_1, x_2 \geq 0$$

Solution: When the two constraints along with the non-negativity conditions are plotted in the $x_1\, x_2$ plane, we have the following figure.

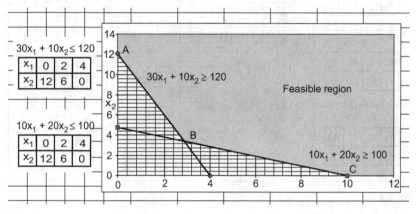

Fig. 2.7

The area inside the region ABC is the feasible region of the problem, and the points A, B, and C denote the extreme points of the region. The coordinates of the extreme points are: A = (0, 12), B = (2.8, 3.6), and C = (10, 0).

The values of the objective function at these points are as follows:

I. at the point A (0, 12), $z = 456$

II. at the point B (2.8, 3.6), $z = \mathbf{206.8}$

III. at the point C (10,0), $z = 250$

Hence, the optimum solution of the problem is $(x_1 = 2.8, x_2 = 3.6)$, and the corresponding minimum value of the objective function is $z^* = 206.8$.

Example 2.8. *Maximize* $z = 4x_1 + x_2$

Subject to the constraints:

$$2x_1 + x_2 \leq 100$$
$$x_1 + x_2 \leq 80$$
$$x_1 \leq 40$$
$$x_1, x_2 \geq 0$$

Solution: When the three constraints along with the non-negativity conditions are plotted in the (x_1, x_2) plane, we have the following figure:

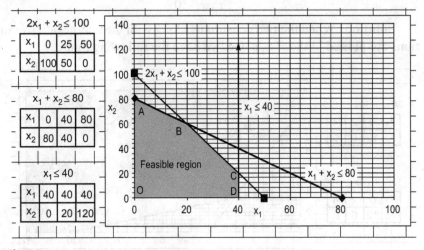

Fig. 2.8

The shaded region OABCD in the previous figure gives the feasible region of the problem, and the values of z at these points (extreme) are as follows:

I. at the point O (0, 0), $z = 0$

II. at the point A (0, 80), $z = \mathbf{80}$

III. at the point B (20, 60), $z = \mathbf{140}$

IV. at the point C (40, 20), $z = \mathbf{180}$

V. at the point D (40, 0), $z = \mathbf{160}$

Clearly $z = \mathbf{180}$ is the maximum value z, and this occurs at the point C (40, 20). Hence, the optimum solution of the problem is ($x_1 = 40$, $x_2 = 20$), and the corresponding maximum value of the objective function is $z^* = 180$.

The following section discusses some special cases in linear programming problems.

2.8. Special Cases

2.8.1. Problem with Multiple Solutions

In some of the linear programming problems, instead of having a unique optimal basic feasible solution to the given problem, we may have an alternative or multiple (infinite) numbers of optimal basic feasible solutions. This case occurs when the line of the objective function is parallel to any one of the constraint lines and lies on one of the edges of the region of feasible solutions, which is illustrated in the following example:

Example 2.9. *Maximize* $z = x_1 + 2x_2$

Subject to the constraints:

$$6x_1 + 4x_2 \leq 24$$
$$x_1 + 2x_2 \leq 6$$
$$-x_1 + x_2 \leq 1$$
$$x_2 \leq 2, x_1, x_2, \geq 0$$

Solution: When the constraints along with the non-negativity conditions are plotted in the $x_1 x_2$ plane, we have the following figure:

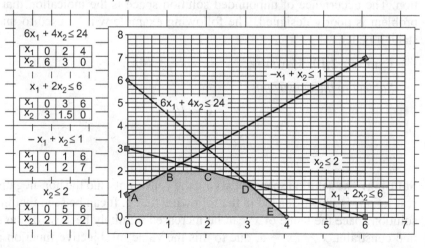

Fig. 2.9

The area inside OABCDE denotes the feasible region of the problem. The values of z at these points are:

    I. at the point O $(0, 0)$, $z = 0$

    II. at the point A $(0, 1)$, $z = 2$

    III. at the point B $(1, 2)$, $z = 5$

    IV. at the point C $(2, 2)$, $z = 6$

    V. at the point D $(3, 1.5)$, $z = 6$

The maximum value of z is 6, and this occurs at two points, C and D. Therefore, it's a case of multiple solutions. All the points which lie on the line segment joining these two points will also give the same value of z.

Note: If a problem has two points giving the same optimum value of the objective function, then this problem will have an infinite number of solutions (all the points on the line segment joining these two points), and all solutions will give the same value of the objective function.

2.8.2. The Problem with Unbounded Solutions

In some problems, the solution space may be unbounded in at least one variable; that is, the value of that variable may be increased indefinitely without violating any of the given constraints, thus giving an infinitely large or small value of the objective function. This kind of problem has an unbounded solution. The occurrence of unbounded solution space is the indication that the problem is poorly designed. The following example will be used to understand it better.

    Example 2.10. *Maximize* $z = 4x_1 + 10x_2$

    Subject to the constraints:

$$x_1 \leq 10$$
$$x_1 + 2x_2 \geq 6$$
$$x_1, x_2 \geq 0$$

Solution: The solution space satisfying the constraints and the non-negativity restrictions is shown shaded in the following figure. It can be seen that in the solution space the value of x_2 can be increased infinitely because of the second constraint, $x_1 + 2x_2 \geq 6$; due to this the value of objective function also can be increased infinitely.

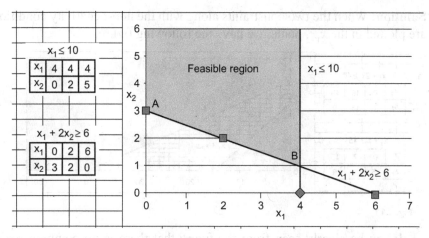

Fig. 2.10

Hence, the value of z can be made arbitrarily large by increasing the value of x_2 indefinitely, which implies that the given LP problem has an unbounded solution.

2.8.3. The Problem with Inconsistent Constraints

The constraints of a given problem are said to be inconsistent if they are mutually exclusive, that is, if no point or solution satisfies all of the constraint equations simultaneously and hence there is no unique common region of feasible solutions. Sometimes a common region of solutions exists but none of the points of this region satisfies the non-negative restrictions of the given LPP. In both these situations, there is no solution for the given problem. Such a problem arises due to wrong or improper formulation of the problem with conflicting constraints.

The problem with inconsistent constraints is said to be an infeasible problem. Infeasibility is purely due to the constraints of the problem the objective function has no role in it.

Example 2.11. *Maximize $z = x_1 + 4x_2$*

Subject to the constraints:

$$x_1 + 2x_2 \geq 40$$
$$x_1 + x_2 \leq 15$$
$$x_1, x_2 \geq 0$$

Solution: When the two constraints along with the non-negativity conditions are plotted in the $x_1 x_2$ plane, we have the following figure:

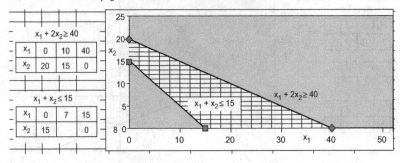

$x_1 + 2x_2 \geq 40$			
x_1	0	10	40
x_2	20	15	0

$x_1 + x_2 \leq 15$			
x_1	0	7	15
x_2	15		0

Fig. 2.11

It can be clearly seen from the figure that there is no common region between the two constraints and hence no feasible solution for the problem. Such a problem is said to be a problem with inconsistent constraints, and the problem does not possess any solution.

2.8.4 The Problem with Redundant Constraint Equations

Any LPP is said to have a redundant constraint if at least one of the constraints does not play any role in finding the optimal solution of the given problem, and if even when it is (or they are) removed from the problem, we get the same solution as was obtained when it was (or they were) present in the problem.

Example 2.12. *Maximize* $z = 2x_1 - x_2$

Subject to the constraints:

$$x_1 + 4x_2 \leq 50$$

$$x_1 + x_2 \leq 20$$

$$1.5x_1 + x_2 \leq 30 \text{ and } x_1, x_2 \geq 0$$

Solution:

The area inside OABC denotes the feasible region of the problem. The values of z at these points are:

I. at the point O (0, 0), $z = 0$

II. at the point A (0, 12.5), $z = -12.5$

III. at the point B (10, 10), $z = 10$

IV. at the point C (20, 0), $z = 30$

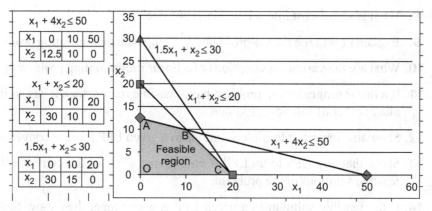

Fig. 2.12

The maximum value of z is 30, and this occurs at the point C. It can be clearly seen that with or without the third constraint, the optimal solution of the problem will remain the same. Hence, the constraint $1.5x_1 + x_2 \leq 30$ is said to be redundant.

Exercises

1. What is a linear programming problem? What are its limitations? Discuss briefly its various areas of application.

2. Write at least five applications of linear programming problems.

3. Explain why solutions to linear programming problems focus on corner points.

4. In the context of a linear programming problem, define the following;
   (*i*) Solution
   (*ii*) Feasible solution
   (*iii*) Basic feasible solution and extreme points of a feasible region
   (*iv*) Optimal basic feasible solution

5. What is a feasible region of a linear programming problem? When is it solved using the graphical method?

6. Give the generalized definition of an LPP.

7. Explain the various approaches to solving an LPP using the graphical method.

8. What is an iso-cost line and an iso-profit line?

9. Explain how to use the iso-profit line in a graphical maximization problem.

10. What are the various assumptions of a linear programming problem?

11. If a linear programming problem has a feasible region, then it has an infinite number of solutions. Explain.

12. Show that the set of feasible solutions to a given LPP forms a convex set.

13. Show that any BFS to an LPP is an extreme point of the convex set of feasible solutions of the problem.

14. If the feasible solution to a given LPP is a polytope, then show that at least one extreme point of the polytope will give the optimum solution of the given LPP.

15. A firm manufactures three products A, B, and C on which the profits earned per unit are $10, $15, and $14, respectively. Each product is processed on two machines M_1 and M_2. Product A requires one minute of processing time on M_1 and three minutes on M_2, B requires two minutes on M_1 and one minute on M_2, while C requires one minute on M_1 and one minute on M_2. Machine M_1 is available for not more that 15 hours, while M_2 is available for 18 hours during any working day. Formulate this problem as an LP model so as to get maximum profit.

16. A shopkeeper plans to sell two types products at costs of $1150 and $1600, The $1150 product yields a profit of $90 and the $1600 product yields a profit of $100. The shopkeeper estimates that the total monthly demand will not exceed 500 units. Find the number of units of each model that should be stocked in order to maximize profit. Also, it is given that the shopkeeper does not want to invest more than $1,000,000 in inventory.

17. A company manufactures two types of models M_1 and M_2. Each M_1 model requires 4 hours of grinding and 2 hours of polishing, while each M_2 model requires 2 hours of grinding and 5 hours of polishing. The company has 2 grinders and 3 polishers. Each grinder works for 40 hours a week and each polisher works for 60 hours a week. Profit on M_1 is $30 and that on a M_2 model is $40. Formulate this as an LPP and solve it graphically. A company manufactures two types of models M_1 and M_2. Each M_1 model requires 4 hours of grinding and 2 hours of polishing, while each M_2 model requires 2 hours of grinding and 5 hours of polishing. The company has 2 grinders and 3 polishers. Each grinder works for

40 hours a week, and each polisher works for 60 hours a week. Profit on M_1 is $30, and that on an M_2 model is $40. Formulate this as an LPP and solve it graphically.

18. The standard weight of a special purpose brick is 5 kg, and it contains two basic ingredients B_1 and B_2. B_1 costs $5/ kg and B_2 costs, $8/ kg. Strength considerations dictate that the brick contains not more than 4 kg of B_1 and a minimum of 2 kg of B_2. Since the demand for the product is likely to be related to the price of the brick, find graphically the minimum cost of the brick satisfying the previous conditions. Further, find the range of the ratio of costs of two materials so that the same solution remains optimal.

19. A company has received a contract to supply gravel to three new construction projects located in towns A, B, and C. The construction engineers have estimated that the required amount of gravel which will be needed at these construction projects are:

Project location	Weekly requirement (truck loads)
A	72
B	102
C	41

The company has 3 gravels pits located in towns X, Y, and Z. The gravel required by the construction projects can be supplied by three pits. The amount of gravel that can be supplied by each pit is as follows:

Plant	X	Y	Z
Available amount (truck loads)	76	82	77

The company has computed the delivery cost from each pit to each project site. These costs (in $) are shown in the following table:

			Location		
			A	B	C
	X		4	8	8
Pit	Y		16	24	16
	Z		8	16	24

Schedule the shipment from each point to each project in such a manner that it minimizes the total transportation cost within the constraints imposed by pit capacities and project requirements. Formulate this transportation problem as a linear programming problem to minimize the total cost of transportation.

20. A company has a scheduling problem. Operators are needed according to the schedule shown in the table. The operators report at the company at the beginning of each period and work for 8 consecutive hours. Let x_j equal the number of operators beginning work in time period j, $j = 1, 2, 3, ..., 6$. Formulate an LP model to hire the minimum number of operators the company needs.

Time period	Operators needed
Midnight to 3 A.M.	5
3 A.M. to 7 A.M.	10
7 A.M. to noon	88
Noon to 3 P.M.	76
3 P.M. to 7 P.M.	50
7 P.M. to midnight	15

21. A factory manufactures three products which are processed through three different production stages. The time required for manufacturing one unit of each of the three products and the daily capacity of the stages are given in the following table:

Stage	Time required per unit			Stage Capacity
	Product 1	Product 2	Product 3	
1	1	1	1	430
2	3	---	2	460
3	1	4	---	420
Profit per unit	3	2	5	

Formulate the previous problem as a linear programming problem.

22. A finished product must weigh exactly 150gms. The two raw materials used in manufacturing the product are A, with a cost of $2 per unit, and

B, with a cost of $8 per unit. At least 14 units of B and not more than 20 units of A must be used. Each unit of A and B weights 5 and 10 grams respectively. How much of each type of raw material should be used for each unit of the final product in order to minimize the cost?

23. A company produces two types of cowboy hats. Each hat of the first type requires twice as much labor time as does each hat of the second type. If all hats are of the second type only, the company can produce a total of 500 hats a day. The market limits daily sales of the first and second types to 150 and 200 hats. Assume that the profit per hat is $8 for type 1 and $5 for type 2. Determine the number of hats of each type to produce to maximize profit.

24. A firm has 240, 370, and 180 kg of wood, plastic, and steel, respectively. The firm produces two products A and B each unit respectively. Each unit of A requires 1, 3, and 2 kg of wood, plastic, and steel, respectively. The corresponding requirement for each unit of B is 3, 4, and 1 kg, respectively. If A sells for $4 and B sells for $6 per unit, then what product mix should the firm produce in order to have maximum gross income? Formulate this as an LPP.

25. Use the corner point approach to solve the following linear programming problems graphically:

(*i*) *Min* $z = 3x_1 + 5x_2$

Subject to

$$x_1 + x_2 \leq 3$$
$$0.5x_1 + 0.5x_2 \leq 1, x_1, x_2 \geq 0$$

(*ii*) *Max* $z = 5x_1 + 3.1 x_2$

Subject to

$$3x_1 + 3.3x_2 \leq 30$$
$$x_2 \leq 6, x_1 \leq 7$$
$$x_1, x_2 \geq 0$$

(*iii*) *Min* $z = 5x_2$

Subject to

$$x_1 + 3.3x_2 \leq 10$$
$$x_1 + 10x_2 \geq 20, x_1, x_2 \geq 0$$

(*iv*) *Max* $z = 1.5x_1 + 1.5x_2$

Subject to

$$x_1 + 2x_2 \leq 10$$

$$3x_1 + 3x_2 \geq 12$$

$$x_2 \leq 3, x_1, x_2 \geq 0$$

(*v*) *Min* $z = -x_1 - x_2$

Subject to

$$-2x_1 + 2x_2 \leq 2$$

$$x_1 - 2x_2 \geq -4 \; x_1, x_2 \geq 0$$

(*vi*) *Max* $z = 5x_1 + x_2$

Subject to

$$3x_1 + 4x_2 \leq 20$$

$$3x_1 + 3x_2 \geq 21$$

$$x_2 \leq 3, x_1, x_2 \geq 0$$

26. Consider the following linear programming formulation:

Min $z = x_1 + 4x_2$

Subject to

$$4x1 + 2x_2 \geq 60$$

$$3x_1 + 10x_2 \geq 120$$

$$x_1 + 1.5x_2 \geq 30$$

and $x_1, x_2 \geq 0$

Graphically illustrate the feasible region and apply the iso-cost line to find the optimal solution.

27. Discuss the following situations using the graphical method:

(*i*) Unbounded solution

(*ii*) Multiple solution

(*iii*) Inconsistent constraint equations

(*iv*) Redundancy

CHAPTER 3

SIMPLEX METHOD-I

3.1. Standard and Canonical Form of the General Linear Programming Problem

The Standard Form

The standard form of a given linear programming problem should have the following four characteristics:

(i) The objective function should be either maximization or minimization.

(ii) All the constraints should be of the equality type.

(iii) The right-hand side vector *b* is non-negative.

(iv) All the decision variables should be greater than or equal to zero.

Definition: The standard form of a general linear programming problem in *m* constraints and *n* variables may be written as follows:

Maximize or Minimize $\quad z = c_1 x_1 + c_2 x_2 + \ldots + c_n x_n$

Subject to the constraints

$$a_{11} x_1 + a_{12} x_2 + \ldots + a_{1n} x_n = b_1$$
$$a_{21} x_1 + a_{22} x_2 + \ldots + a_{2n} x_n = b_2$$
$$\vdots \qquad \vdots \qquad \vdots$$
$$a_{i1} x_1 + a_{i2} x_2 + \ldots + a_{in} x_n = b_i$$
$$\vdots \qquad \vdots \qquad \vdots$$
$$a_{m1} x_1 + a_{m2} x_2 + \ldots + a_{mn} x_n = b_m$$
$$x_1, x_2, \ldots x_n \geq 0$$

...(3.1)

So, the standard form of a given linear programming problem consists of solving a set of simultaneous linear equations such that the solution satisfies the non-negativity conditions and optimizes the linear objective function

The following is the matrix notation of the previous problem:

Maximize or Minimize z = cx

Subject to the constraints

$$Ax = b$$

and $$\qquad x \geq 0 \qquad \qquad ...(3.2)$$

Here c is a $1 \times n$ unit profit (or cost) vector, x is an $n \times 1$ decision variables vector, A is an $m \times n$ coefficient matrix, and b is a non-negative m-dimensional column (resource) vector.

The Canonical Form

In matrix notation, the canonical form can be written:

Maximize or Minimize $\quad z = c_{NB} x_{NB}$

Subject to the constraints

$$A_1 x_{NB} + I_m x_B = b$$

and $$\qquad x_{NB}, x_B \geq 0 \qquad \qquad ...(3.3)$$

where c_{NB} is the $(n - m)$ components cost coefficients vector associated with non-basic variables x_{NB} in the objective function; A_1 is the $m \times (n - m)$ order coefficient matrix of the non-basic variables in the constraint equations; I_m is an $m \times m$ order identity matrix associated with the basic variables x_B and b is a non-negative m-dimensional column vector. The decision variables x_{NB} and x_B are $(n - m)$ and m-dimensional column vectors, respectively. The basic solution $x_B = b$ of the canonical form is obtained by setting $x_{NB} = 0$. The basic solution is feasible because $b \geq 0$. The variables in the vector x_{NB} are called **non-basic variables** and in the vector x_B are called **basic variables.**

Note: In the canonical form, the following conditions should be true:

(i)   The objective function should be either in maximization or minimization.

(ii) All the constraints should be of the equality type.

(iii) All the decision variables should be greater than or equal to zero.

(iv) The right-hand side vector b is non-negative.

(v)   *The objective function should be purely in terms of non-basic variables.*

(vi)   *The total number of basic variables should be equal to the number of constraints.*

(vii) *Each constraint should have exactly one basic variable with coefficient 1.*

Now, it is clear that the canonical form of a given LPP will also represent a standard form of the same problem, whereas the opposite does not always hold true —which means a standard form may or may not represent a canonical form.

3.2. Slack and Surplus Variables

Slack Variable: The non-negative variable which is added to the LHS of the constraint with a less than or equal to inequality ($\leq$) relation, to convert it into an equality (=), is called a slack variable.

For example, consider $2x_1 + 3x_2 \leq 60$ as one of the constraints of any LPP. Adding s_1 (≥ 0) to the left-hand side of the constraint, we have

$$2x_1 + 3x_2 + s_1 = 60$$

Here, the non-negative variable s_1 is a slack variable.

Surplus Variable: The non-negative variable which is subtracted from the left-hand side of a greater than or equal to inequality ($\geq$) constraint, to convert it into an equality (=), is called a surplus variable.

For example, consider $2x_1 + 3x_2 \geq 60$ as one of the constraints of any LPP. Subtracting s_1 (≥ 0) from the left-hand side of the constraint, we have

$$2x_1 + 3x_2 - s_1 = 60$$

Here, the non-negative variable s_1 is a surplus variable.

Note: The prices assigned to slack and surplus variables in the objective function will always be zero because these variables are used only to convert inequalities into equations; otherwise, they do not play any role in optimizing the objective function.

Unrestricted Variable: A decision variable is said to be unrestricted in sign if it is allowed to take any value, such as positive, negative, or zero. In order to solve any LPP containing unrestricted variables, there is a need to convert the problem into the form which contains only non-negative variables.

For example, let x_j be a variable which is defined to be unrestricted in sign. Thus, we can write:

$$x_j = x'_j - x''_j \qquad \text{where} \quad x'_j, x''_j \geq 0$$

The variable x_j can be written as the difference of two non-negative variables x_j' and x_j'', such that: (*i*) if $x_j' > x_j''$ then $x_j > 0$, (*ii*) if $x_j' < x_j''$ then $x_j < 0$, and (*iii*) if $x_j' < x_j''$ then $x_j = 0$.

Note: After converting any problem having variables unrestricted in sign into a standard linear programming problem having all non-negative variables, if we solve this problem, then only either one of the two variables x_j' and x_j'', or either of them, will appear in the final solution. They cannot both simultaneously be present in the optimal basic feasible solution since vectors corresponding to these two variables are linearly dependent and cannot appear in the basis at the same time. The variable which will be present in the optimal solution at the positive level will determine the sign of the original variable x_j.

Example 3.1. *Write the standard form of the following LPP:*

Maximize $$z = x_1 + 2x_2 - x_3$$

Subject to the constraints

$$x_1 + x_2 + x_3 \leq 40$$
$$x_1 + 2x_2 \leq 26$$
$$x_1 + x_2 + 2x_3 \leq 36$$

and $$x_1, x_2, x_3 \geq 0$$

Solution: Introducing the slack variables s_1, s_2, and s_3 in the first, second, and third constraints, respectively, we have

$$x_1 + x_2 + x_3 + s_1 = 40$$
$$x_1 + 2x_2 + s_2 = 26$$
$$x_1 + x_2 + 2x_3 + s_3 = 36$$

So, the standard form of the given problem is

Maximize $$z = x_1 + 2x_2 - x_3$$

Subject to the constraints

$$x_1 + x_2 + x_3 + s_1 = 40$$
$$x_1 + 2x_2 + s_2 = 26$$
$$x_1 + x_2 + 2x_3 + s_3 = 36$$

and $$x_1, x_2, x_3, s_1, s_2, s_3 \geq 0$$

Example 3.2. *Put the following problem into its standard form:*

Maximize $\qquad\qquad\qquad\qquad z = 6x_1 + x_2$

Subject to the constraints

$$x_1 + x_2 \geq 4$$
$$x_1 + x_2 \leq 6$$
$$x_1 \geq 0 \text{ and } x_2 \text{ is unrestricted in sign}$$

Solution: Introducing the surplus variable $s_1 \geq 0$ to the first constraint and the slack variable $s_2 \geq 0$ in the second constraint, we have

$$x_1 + x_2 - s_1 = 4$$
$$x_1 + x_2 + s_2 = 6$$

Since x_2 is unrestricted in sign, we can write

$$x_2 = x_2' - x_2'' \text{ where } x_2', x_2'' \geq 0$$

So, the standard form of the given problem is

Maximize $\qquad\qquad\qquad\qquad z = 6x_1 + x_2' - x_2''$

Subject to the constraints

$$x_1 + x_2' - x_2'' - s_1 = 4$$
$$x_1 + x_2' - x_2'' + s_2 = 6$$
$$x_1, x_2', x_2'', s_1, s_2 \geq 0$$

Example 3.3. *Put the following problem into its canonical form:*

Minimize $z = x_1 - x_2 + 3x_3 - x_4$

Subject to the constraints

$$x_1 + x_2 + x_3 + 3x_4 = 14$$
$$x_2 + 3x^3 + x_4 = 10$$
$$x_1, x_2, x_3, x_4 \geq 0$$

Solution: Since the problem has two constraints, it will have two basic variables. Let $x_B = (x_1, x_2)$ and $x_{NB} = (x_3, x_4)$ be the vectors of basic and non-basic variables of the problem.

Putting the value of x_2 obtained from the second constraint, the first constraint can be written as:

$$x_1 + (10 - 3x_3 - x_4) + x_3 + 3x_4 = 14$$
$$\Rightarrow \qquad x_1 - 2x_3 + 2x_4 = 4$$

Also expressing the objective function purely in terms of the non-basic variables, we have:

$$z = (4 + 2x_3 - 2x_4) - (10 - 3x_3 - x_4) + 3x_3 - x_4$$
$$\Rightarrow \qquad z = -6 + 8x_3 - 2x_4$$

Now we have

Minimize $\qquad z = -6 + 8x_3 - 2x_4$

Subject to the constraints

$$x_1 - 2x_3 + 2x_4 = 4$$
$$x_2 + 3x_3 + x_4 = 10$$
$$x_1, x_2, x_3, x_4 \geq 0$$

The previous problem is the canonical expression of the given problem with x_1 and x_2 as basic variables.

3.3. Algebraic Simplex Method

The simplex method is a very powerful method developed by George Dantzig for solving any given linear programming problem. It is an iterative method to solve any LPP, either exactly in a finite number of steps or by giving an indication that the problem has an unbounded solution or no solution at all. As we have already discussed, the common feasible region of any given LPP always forms a convex set, and one of the extreme points (basic feasible solution) of this bounded convex set will always be an optimal solution. Since there are only finite numbers of extreme points of any bounded region, the simplex method moves from one extreme point to another adjacent extreme point, after starting from the initial extreme point, and ultimately reaches to the optimal extreme point in a finite number of steps.

Out of all adjacent extreme points, the simplex method selects that extreme point giving the maximum improvement in the value of the objective function; if no further improvement is possible, the current extreme point is optimal. At any stage, if the simplex method reaches to an extreme point which has an edge leading to infinity and the objective function can be improved infinitely by moving along that edge, then it gives an indication that the problem has an unbounded solution. Moving from one corner point to another in the simplex method is the same as moving from one basic feasible solution to another by changing one basic variable in the given basis by a non-basic variable so that we get a new basis and hence a new basic feasible solution.

In this section, we will discuss the **algebraic approach** to this method. Here, the basic idea behind this technique is explored using an example so as to make the reader understand the method in a better way.

Example 3.4. *Max z $=3x_1 + 1.5x_2 + 2x_3 + 2x_4$*

Subject to the constraints

$$x_1 + x_2 + x_3 + 2x_4 = 60$$
$$x_2 + 3x_3 + 3x_4 = 26$$
$$x_1, x_2, x_3, x_4 \geq 0$$

Solution: Since there are two constraints in the given problem, we should have two basic variables. The simplex calculation will be started with the canonical form of the problem. To convert the given problem into its canonical form, we arbitrarily select any two variables as basic variables, which will provide us with the initial basic feasible solution of the problem. Let $x_B = (x_1, x_2)$ and $x_{NB} = (x_3, x_4)$ be the initial vectors of basic and non-basic variables of the problem. Using elementary row operations, we can write the constraints and z equation in the canonical form as

$$x_1 + (26 - 3x_3 - 3x_4) + x_3 + 2x_4 = 60$$
$$\Rightarrow \qquad\qquad x_1 - 2x_3 - x_4 = 34 \qquad\qquad ...(3.4)$$
$$x_2 + 3x_3 + 3x_4 = 26 \qquad\qquad ...(3.5)$$

Replacing the values of x_1 and x_2 from (3.4) and (3.5) in the objective function, we have

$$z = 3 (34 + 2x_3 + x_4) + 1.5 (26 - 3x_3 - 3x_4) + 2x_3 + 2x_4$$
$$= 141 + 3.5x_3 + 0.5x_4$$

So, the canonical form of the given problem is

Maximize z

Subject to

$$x_1 - 2x_3 - x_4 = 34 \quad \text{...(3.6)}$$

$$x_2 + 3x_3 + 3x_4 = 26 \quad \text{...(3.7)}$$

$$z - 3.5x_3 - 0.5x_4 = 141 \quad \text{...(3.8)}$$

$$x_1, x_2, x_3, x_4 \geq 0$$

The value of the objective function at the associated basic feasible solution **(34, 26, 0, 0)** is **141**. Now the key behind the simplex method is to move from one basic feasible solution to another basic feasible solution that gives a better value of the objective function, that is, the larger value of z in this case. This is done by replacing one variable in the basis with one non-basic variable. On carefully observing the z equation given in (3.8), it can be seen that the value of z can be improved if at least one of the coefficients of the non-basic variable is negative (**this will be called the optimality condition**). If more than one variable has a negative coefficient, then the variable with the most negative coefficient will enter the basis (the condition for deciding about entering a variable into the basis) in the next calculation. In this case, x_3 will enter the basis.

Now, x_3 will replace either x_1 or x_2 in the basis. To decide about this, we set the non-basic variable x_4 equal to zero in equations (3.6) and (3.7) and we get

$$x_1 - 2x_3 = 34 \quad \text{...(3.9)}$$

$$x_2 + 3x_3 = 26 \quad \text{...(3.10)}$$

Solving for x_1 and x_2 gives

$$x_1 = 34 + 2x_3 \quad \text{...(3.11)}$$

$$x_2 = 26 - 3x_3 \quad \text{...(3.12)}$$

Since x_1 and x_2 are both non-negative, we have

$$0 \leq 34 + 2x_3 \quad \Rightarrow \quad -17 \leq x_3$$

And

$$0 \leq 26 - 3x_3 \quad \Rightarrow \quad x_3 \leq \frac{26}{3}$$

Clearly, x_3 cannot be made arbitrarily large as it has to satisfy both inequalities. We are here to find the maximum possible value of x_3 which will satisfy both inequalities. Since the first inequality does not have an upper bound on x_3, the upper limit for x_3 is solely determined by the second equality and is equal to $\frac{26}{3}$. And hence, x_3 will replace x_2 in the basis.

(These are the criteria needed to decide about the leaving variable from the basis in the simplex calculation. Calculating $\min\left(\dfrac{b_i}{a_{ij}}, a_{ij} > 0\right)$ for the constraint equations of the present basic variables and the corresponding variable for which the minimum ratio exists results in the leaving variable from the basis. Here, a_{ij} is the coefficient of the new basic variable x_j in the constraint equation of the ith basic variable x_i, and b_i is the present value of x_i).

Now, $x_B = (x_1, x_3)$ and $x_{NB} = (x_2, x_4)$. Letting $x_3 = \dfrac{26}{3}$ gives the basic feasible solution $\left(\dfrac{150}{3}, 0, \dfrac{26}{3}, 0\right)$, and the value of z at this point is $\dfrac{514}{3}$. Rewriting (3.6), (3.7), and (3.8) in the canonical form on the basis of the new basic and non-basic variables, we have

$$x_1 + \frac{2}{3}x_2 + x_4 = \frac{154}{3} \qquad \qquad \text{...(3.13)}$$

$$\frac{1}{3}x_2 + x_3 + x_4 = \frac{26}{3} \qquad \qquad \text{...(3.14)}$$

$$z + \frac{3.5}{2}x_2 + 3x_4 = \frac{514}{3} \qquad \qquad \text{...(3.15)}$$

$$x_1, x_2, x_3, x_4 \leq 0$$

Since the coefficient of the non-basic variables in the z-row in (3.15) are all positive, no further improvement on the value of z is possible; hence, the present basic feasible solution is the optimal basic feasible solution of the given problem, and the corresponding value of z is the maximum value of z.

Optimum solution: $x_1 = \dfrac{154}{3}$, $x_2 = 0$, $x_3 = \dfrac{26}{3}$, $x_4 = 0$, and the maximum value of z is $\dfrac{514}{3}$.

Example 3.5. *Maximize $z = 4x_1 + x_2$*

Subject to the constraints:

$$2x_1 + x_2 \leq 100$$

$$x_1 + x_2 \leq 80$$

$$x_1 \leq 40$$

$$x_1, x_2 \geq 0$$

Solution: Writing the given problem into its canonical form by introducing the slack variables s_1, s_2, and s_3, we have

Maximize z

Subject to the constraints

$$2x_1 + x_2 + s_1 = 100 \qquad \ldots(3.16)$$

$$x_1 + x_2 + s_2 = 80 \qquad \ldots(3.17)$$

$$x_1 + s_3 = 40 \qquad \ldots(3.18)$$

$$z - 4x_1 - x_2 = 0 \qquad \ldots(3.19)$$

$$x_1, x_2, x_1, s_2, s_3 \geq 0$$

Let $x_B = (s_1, s_2, s_3)$ and $x_{NB} = (x_1, x_2)$. The value of z at this point is zero. Clearly, from (3.19), we can see that x_1 will enter the basis. To decide on the departing variable, we put $x_2 = 0$ in (3.16), (3.17), and (3.18), and then we have

$$s_1 = 100 - 2x_1 \geq 0 \quad \Rightarrow \quad x_1 \leq 50 \qquad \ldots(3.20)$$

$$s_2 = 80 - x_1 \geq 0 \quad \Rightarrow \quad x_1 \leq 80 \qquad \ldots(3.21)$$

$$s_3 = 40 - x_1 \geq 0 \quad \Rightarrow \quad x_1 \leq 40 \qquad \ldots(3.22)$$

The maximum possible value of x_1 which satisfies (3.20)–(3.22) is 40, and this corresponds to (3.22), and hence x_1 will replace s_3 in the basis. Letting $x_1 = 40$ gives the basic feasible solution (40, 0, 20, 40, 0), and the value of z at this point is 160.

Rewriting (3.16)–(3.19) in the canonical form on the basis of the new basic and non-basic variables, we have

$$x_2 + s_1 - 2s_3 = 20 \qquad \ldots(3.23)$$

$$x_2 + s_2 - s_3 = 40 \qquad \ldots(3.24)$$

$$x_1 + s_3 = 40 \qquad \ldots(3.25)$$

$$z + 4s_3 - x_2 = 160 \qquad \ldots(3.26)$$

Since the coefficient of x_2 in (3.26) is negative, x_2 enters the basis. Setting $s_3 = 0$ in (3.23)–(3.25), we get

$$s_1 = 20 - x_2 \geq 0 \quad \Rightarrow \quad x_2 \leq 20 \qquad \ldots(3.27)$$

$$s_2 = 40 - x_2 \geq 0 \quad \Rightarrow \quad x_2 \leq 40 \qquad \ldots(3.28)$$

$$x_1 = 40 \geq 0 \qquad \ldots(3.29)$$

Again the value of x_2 must satisfy both (3.27) and (3.28); the maximum possible value of x_2 is 20, and this corresponds to (3.27), and thus x_2 will replace s_1 in the basis. Letting $x_1 = 40$ and $x_2 = 20$ gives the basic feasible solution (40, 20, 0, 20, 0), and the value of z at this point is 180.

Rewriting the z-row in terms of the new non-basic variables, we have:

$$z + 2s_3 + s_2 = 180 \qquad \qquad ...(3.30)$$

Since the coefficients of the non-basic variables in the z-row in (3.30) are all positive, no further improvement on the value of z is possible. Therefore, the present basic feasible solution is the optimal basic feasible solution of the given problem, and the corresponding value of z is at its maximum.

Example 3.6. *Maximize $z = 4x_1 + 10x_2$*

Subject to the constraints:

$$x_1 + x_3 = 10$$

$$x_1 + 2x_2 - x_4 = 6$$

$$x_1, x_2, x_3, x_4 \geq 0$$

Solution: Let $x_B = (x_2, x_3)$ and $x_{NB} = (x_1, x_4)$ be the vectors of the initial basic and non-basic variables of the problem. Using elementary row operations, we can write the constraints and z equation in the canonical form as:

Max z

Subject to the constraints

$$x_1 + x_3 = 10 \qquad \qquad ...(3.31)$$

$$\frac{1}{2}x_1 + x_2 - \frac{1}{2}x_4 = 3 \qquad \qquad ...(3.32)$$

$$z + x_1 - 5x_4 = 30 \qquad \qquad ...(3.33)$$

$$x_1, x_2, x_3, x_4 \geq 0$$

Clearly, x_4 will enter the basis in the next step, as it is the only non-basic variable with a negative coefficient in (3.33). Now, x_4 will replace either x_2 or x_3 in the basis. To decide this, we set non-basic variable (x_1) equal to zero in equations (3.31) and (3.32), and we have

$$x_3 = 10 \qquad \qquad ...(3.34)$$

$$x_2 - \frac{1}{2}x_4 = 3 \quad \Rightarrow \quad x_2 = 3 + \frac{1}{2}x_4 \qquad \qquad ...(3.35)$$

From (3.35), we can see that the value of x_4 can be increased as much as we wish, so as to have a very large value of x_2 and thus the large value of z. This is a situation where the problem has an unbounded solution. And hence the given problem has an unbounded solution with $x_1 = 0$, $x_3 = 10$, $x_4 = 0$, and an infinitely large value of x_2.

Algorithm of the Algebraic Simplex Method: The steps involved in the previous calculations may be summarized as follows:

1. First convert the given problem into its canonical form, where all the inequalities are converted into equations by adding slack and surplus variables so that all the constraints are in equation form, the R.H.S. of the equations are positive (if it is negative then multiply both sides of the equation by a negative sign), and the objective function is in maximization (minimization).

2. Obtain the **initial basic feasible solution** to start the calculation where the coefficient matrix of the initial basic variables should form an identity matrix. The initial basic feasible solution is given by $X_B = B^{-1} b \geq 0$, where B is a basis matrix, b is an RHS vector, and X_B is a vector containing basic variables. Initially, B is an identity matrix, and hence $X_B = b \geq 0$ gives the solution of initial basic variables. Write the objective function purely in terms of the non-basic variables.

3. Check the **optimality condition** of the present basic feasible solution. We look at the coefficients of the objective function (z) equation to find one that is negative (positive). If there is none, we have an optimal solution. Otherwise, let the variable with the most negative (positive) value (say x_r) in the z equation enter the basis.

4. Putting the values of other non-basic variables equal to zero in the constraint equations, we shall now increase the value of x_r until the first basic variable x_t becomes zero. If this never happens, that is, if no basic variable starts having negative value, however large x_r may be, then the problem will have an unbounded solution. Otherwise, we will replace x_t by x_r in the basis.

5. Change the basis by replacing the departing basic variable with the new entering non-basic variable. Obtain the new basic feasible solution. The value of z so obtained will have a better value than the previous one.

6. Repeat the previous steps until no further improvement is possible. The whole process will terminate after a finite number of steps with one of the following conclusions:

(*a*) the problem has no feasible solution;

(*b*) the problem has a finite optimal solution; or

(*c*) the problem has an unbounded solution.

Note. The value of z improves at each iteration unless there are constraints with zero value. In this case, the new value of z is same as its old value. This situation is called degeneracy, and such a case needs to be handled differently. For the time being, we assume that we do not have this situation.

3.4. Relationship between the Simplex and Graphical Methods

Now that it can be seen that the simplex method is similar to solving a system of linear equations. In fact, this method does solve a system of linear equations (to be discussed in next chapter) and the solution is derived. Further, the method not only solves the equations but also optimizes the objective function of the problem. There are a number of methods for solving simultaneous linear equations. For example, the Gauss-Jordan method discussed in Chapter 1 is one such method which has a close relation to the simplex calculation.

The simplex method as seen previously is an iterative method of solving a given linear programming problem. The method starts its calculation with an initial basic feasible solution and then repeats the solution process by removing one basic variable from the basis and allowing another from the non-basic variables to enter the basis, making successive improvements until the optimal solution is found. It is sometimes referred to as an **adjacent extreme point solution procedure** because it generally begins at a feasible extreme point and then successively evaluates the adjacent extreme point until one representing the optimal solution is found. It may be recalled that in the graphical method the optimum solution (if it exists) of a problem is found at one of the extreme points. To understand this more deeply, we consider Example 3.5 of the previous section:

Maximize $\qquad\qquad z = 4x_1 + x_2$

Subject to the constraints:

$$2x_1 + x_2 \leq 100$$
$$x_1 + x_2 \leq 80$$
$$x_1 \leq 40 \quad \text{and} \quad x_1, x_2 \geq 0$$

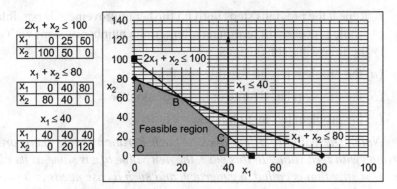

Fig. 3.1

Let the simplex method start its calculation at the origin $(0, 0)$, which means at this point x_1 and x_2 are both non-basic variables. Now, since in the objective function equation x_1 has a larger coefficient than x_2, we shall allow x_1 to enter the basis, so we move to the adjacent point D $(40, 0)$, and finally, it arrives at the extreme point B $(40, 20)$, which is an optimal solution.

To use the simplex method to solve the previous problem, the constraints need to be converted into canonical form.

$$2x_1 + x_2 + s_1 = 100$$
$$x_1 + x_2 + s_2 = 80$$
$$x_1 + s_3 = 40$$

Since there are three equations in five variables, potentially a maximum of $(5!/3!\, 2!) = 10$ basic solutions are possible for this problem. However, only five of these are basic feasible solutions, which correspond to five extreme points of the feasible region. The solution values of all five variables associated with each extreme point are shown in the following table.

Extreme Point (in Fig. 3.1)	x_1	x_2	s_1	s_2	s_3	z
O	0	0	100	80	40	0
A	0	80	20	0	40	80
B	20	60	0	0	20	140
C	40	20	0	20	0	180
D	40	0	20	40	0	160

It can also be easily noted that the values of all the variables, including the slack variables, are non-negative.

3.5. Simplex Method in Tabular Form

Though easy, the graphical method has its own limitations. This method is ideal for problems with only two variables. Whereas the algebraic form of the simplex method may be best for understanding the underlying concept of the simplex algorithm, it's not always convenient to solve a linear programming problem in this way; so, a better way of finding a solution needs to be explored. When the problem is solved manually, the tabular form of solving a linear programming problem will always be a more convenient approach. The tabular form of the simplex method records only the essential information, namely, the coefficient of the variables, the values on the right-hand side, and the basic variables. This not only saves the computation time but also reduces complication in its approach.

To solve a linear programming problem with a maximization (minimization) objective function using the simplex method in tabular form, one can use the following steps:

(i) Write the given problem into its canonical form.

(ii) Create the initial simplex tableau using the initial basic feasible solution.

(iii) **Entering Rule:** Locate the most negative (positive) entry in the bottom row (or z-row). The column corresponding to this entry is called the entering column or pivot column (in a case of a tie, choose any column arbitrarily). The non-basic variable corresponding to this pivot column is the entering variable in the basis.

(iv) **Leaving Rule:** Compute the ratio of the entries in the *b*-Column (solution column) with the corresponding positive entries in the entering column, that is, compute $\dfrac{b_i}{a_{ij}}$, $a_{ij} > 0$, where j is the entering column. Now the departing basic variable corresponds to $\min\left(\dfrac{b_i}{a_{ij}}, a_{ij} > 0\right)$ and the corresponding row is called the pivot row. If all entries in the entering column are ≤ 0, then the problem will have an **unbounded solution**. The entry at the crossing point of the pivot column and pivot row is called the pivot element.

(v) Using elementary row operations convert the pivot element into 1 and all other entries in the entering column as 0. This process is called **pivoting.**

(vi) If all entries in the bottom row are zero or positive (negative), then this is the final tableau and the condition is called the **optimality condition,** which is used to determine whether there is any chance of improvement in the value of the objective function. If the condition is not satisfied, then go to step *(iii)* and repeat the same procedure; otherwise, stop. As in the case of the algebraic method, the whole process will terminate after a finite number of steps with one of the following conclusions:

(a) the problem has no feasible solution;

(b) the problem has a finite optimal solution; or

(c) the problem has an unbounded solution.

This algorithm will be better explained using the following example.

Example 3.7. *Max z = 40x_1 + 35x_2*

Subject to the constraints

$$2x_1 + 3x_2 \le 60$$

$$4x_1 + 3x_2 \le 96$$

$$x_1, x_2 \ge 0$$

Solution: Introducing the slack variables s_1 and s_2, the standard form of the given problem is:

$$Max\ z = 40x_1 + 35x_2$$

Subject to the constraints:

$$2x_1 + 3x_2 + s_1 = 60$$

$$4x_1 + 3x_2 + s_2 = 96$$

$$x_1, x_2, s_1, s_2 \ge 0$$

Initialization: This method starts with the canonical form of the given problem. The initial basic variable, whose coefficient matrix gives an identity matrix, will be identified. Now, the problem can be rewritten as:

$$2x_1 + 3x_2 + s_1 = 60$$

$$4x_1 + 3x_2 + s_2 = 96 \qquad \text{...(3.36)}$$

$$z - 40x_1 - 35x_2 = 0$$

$$x_1, x_2, s_1, s_2 \ge 0$$

For starting the simplex calculation, we need to have an identity matrix of size 2×2, and the columns corresponding to s_1 and s_2 in the constraint equations of (3.36) are going to provide us with the identity matrix. Thus, $x_B = (s_1, s_2)$ and $x_{NB} = (x_1, x_2)$.

Initial Table

Basic Variables	z	x_1	x_2	s_1	s_2	Solution
s_1	0	a_{11}	a_{12}			b_1
s_2	0	a_{21}	a_{22}	B^{-1}		b_2
z	1	$-c_1$	$-c_2$	$-c_3$	$-c_4$	0

The z column plays no role in finding the optimal solution, and hence we will not be including this column in any of the tables in the remaining calculations.

Basic Variables	x_1	x_2	s_1	s_2	Solution	Ratio (Solution/a_{i1} (> 0))	
s_1	2	3	1	0	60	$\dfrac{60}{2} = 30$	
s_2	4	3	0	1	96	$\dfrac{96}{4} = 24$	Pivot Row
z	−40	−35	0	0	0		
	Pivot Column						

Optimality Test: An optimality check will be done on the current basic feasible solution (BFS). The current BFS will be optimal if all the elements in the z-row are $\geq$ ($\leq$) in the case of maximization (minimization), or else we move to the next table.

In the previous table, since the elements in the z row are not all greater than or equal to zero (≥ 0), the optimality condition is not satisfied, and hence we move to the next table by **allowing a non-basic variable to enter the basis and departing a basic variable from the basis at the same time.**

It can be seen that the **most negative element in the z row is -40**, so x_1 enters the basis in the next iteration, and this column corresponding to x_1 (the entering variable) is called the **pivot column.**

To decide on the leaving variable, we find out **the minimum ratio between the elements of the solution column and pivot column (which are strictly greater than zero).** As can be seen from the previous table, the minimum ratio is 24 and this corresponds to s_2. Therefore, s_2 now will be the leaving variable, and this row will be called the **pivot row.** The point of intersection of the pivot row and pivot column is called the **pivot element, which in the previous table is 4.** Therefore, our new basic and non-basic variables are:

$$X_B = (s_1, x_1) \text{ and } X_{NB} = (x_2, s_2)$$

To construct the new table, we shall use the concept of the Gauss-Jordan elimination method discussed in Chapter 1, which includes the following steps:

I. For the row of the new basic variable (in this case x_1), that is, the pivot row:

   (a) **Replace the leaving variable by the entering variable in the basic variables column.**

   (b) **Elements of the new pivot row = current pivot row ÷ pivot element, that is, the current pivot row/pivot element.**

II. All the remaining rows of the new table, including the z-row, will be computed as:

New row = current row – (pivot column coefficient of the current row)*

(Row of the new basic variable)

1. The row of the new basic variable (in this case x_1) = $\frac{1}{4}$ (4 3 0 1 9 6)

Basic variables	x_1	x_2	s_1	s_2	Solution
s_1					
x_1	1	0.75	0	0.25	24

2. The new row for s_1 = (2 3 1 0 60) – 2 * (1 0.75 0 0.25 24)

Basic Variables	x_1	x_2	s_1	s_2	Solution
s_1	0	1.5	1	–0.5	12
x_1	1	0.75	0	0.25	24
z					

3. The new row for $z = (-40\ -35\ 0\ 0\ 0) - \{(-40) * (1\ 0.75\ 0\ 0.25\ 24)\}$

Complete First Iteration Table:

Basic variables	x_1	x_2	s_1	s_2	Solution	Ratio
s_1	0	1.5	1	-0.5	12	8
x_1	1	0.75	0	0.25	24	32
z	0	−5	0	10	960	

Clearly, x_2 replaces s_1 in the basis in the next table.

$\dfrac{1}{1.5}$ (Pivot row of the previous table)

Old row of $x_1 - 0.75*$ row of new basic variable (x_2).

Old row of $z - (-5)*$ row of new basic variable (x_2).

Second Iteration:

Basic variables	x_1	x_2	s_1	s_2	Solution
x_2	0	1	0.67	−0.33	8
x_1	1	0	−0.5	0.5	18
z	0	0	3.33	8.33	1,000

Since all the elements in the z-row are ≥ 0, the optimality condition is now satisfied. Hence, the optimum solution of the problem is $x_1 = 18$ and $x_2 = 8$, and the corresponding maximum value of z is **1,000**.

3.6. Use of Solver in MS-Excel for Solving a Linear Programming Problem

The stepwise solution of the problem given in Example 3.7 using the in-built "Solver" module of MS-Excel 2007 has been demonstrated in this section.

Step 1. Enter your data as shown in the following figure.

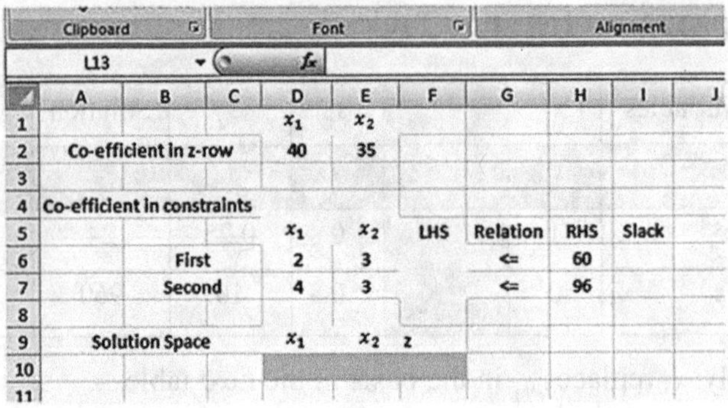

Fig. 3.2

Step 2. Define the LHS, Slack, and z value.

First LHS = Sum product of the first constraint coefficient and the optimum value of x_1 and x_2 in the solution space, which can be obtained when the problem gets solved. Initially, we can enter any value or the computer will assume it to be when none is entered.

Second LHS = Sum product of the second constraint coefficient and the optimum value of x_1 and x_2 in the solution space, which we will have after solving the problem.

Slack = RHS − LHS

z = Sum product of the coefficients of the decision variables in the objective function and their optimum value in the solution space.

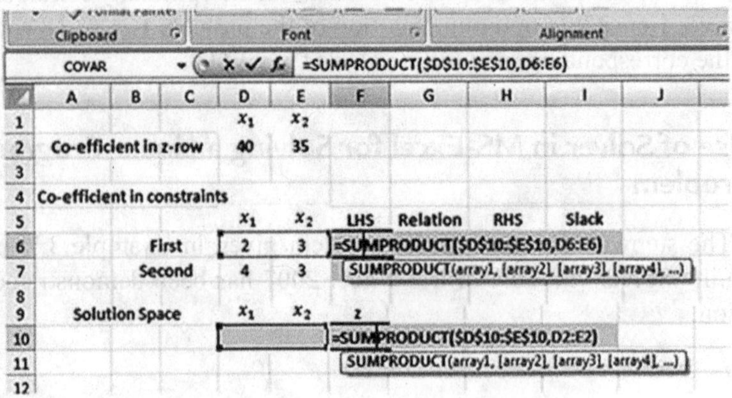

Fig. 3.3

Step 3. Go to Data menu option in EXCEL and select Solver (on the top right corner).

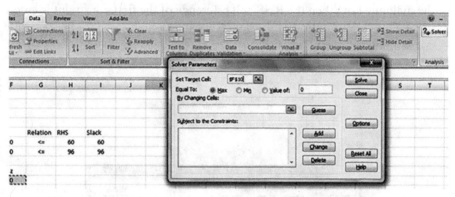

Fig. 3.4

Step 4. Fill up the cells in the Solver parameters window.

In the **Set Target Cell**, enter the cell address from the spread sheet where the value of z (where the **Sumproduct** formula was written) is calculated. Define the objective of the problem by selecting **Max or Min.**

In the **By Changing Cells window** select the cells in the solution space where the values of x_1 and x_2 are to be displayed. And then click **Add** in the **Subject to the constraints** option.

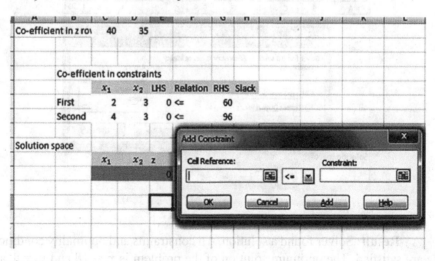

Fig. 3.5

In the **Add Constraint** window, select the call address of the LHS for the first constraint from the worksheet and enter it in the **Cell reference** option of the dialogue box; then the RHS cell reference of the first constraint is entered in the **Constraint** option of the box. Repeat the same steps for the remaining constraints, and then click **OK**.

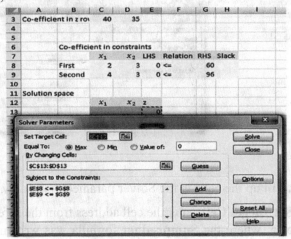

Fig. 3.6

Step 5. Now Click Solve to have the final solution.

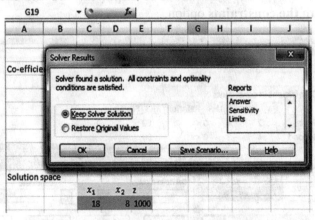

Fig. 3.7

Result: Solver found a solution. All constraints and optimality conditions are satisfied. The optimum solution of the problem is $x_1 = 18$ and $x_2 = 8$, and the corresponding maximum value of z is **1,000**.

3.7. Use of Jensen Add-Ins for a Linear Programming Problem

Paul Jensen, Professor Emeritus at the University of Texas-Austin, used the Internet to share his experiences with Excel-based add-ins for computing various Operational Research (O.R.) problems. On the site he describes models for an enormous variety of O.R. problems and has a collection of Microsoft Excel add-ins that implement O.R. methods. All these add-ins are available for free and can be easily downloaded.

The following steps explain how the Jensen add-ins can be used to solve the same Example (3, 7):

Step 1. Download the Jensen Add-In from *http://www.me.utexas. edu/~jensen/ORMM/ excel/library_windows.html* and install the add-in files *lpip_solver.xla* and *mp_models.xla*.

Step 2. Go to Add-Ins and then to OR_MM and select **_Linear/ Integer**.

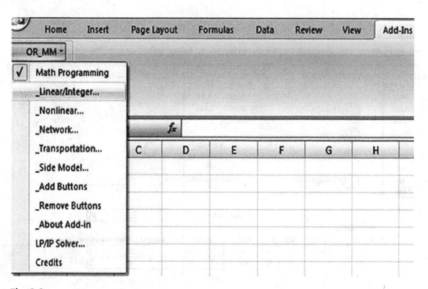

Fig. 3.8

Step 3. In the linear programming dialog box, define the objective of the problem, the number of variables, and the number of constraints. Also select "none" in the integer variable option box, select "Include Minimums" and "Jensen LP/IP," and then click **OK**.

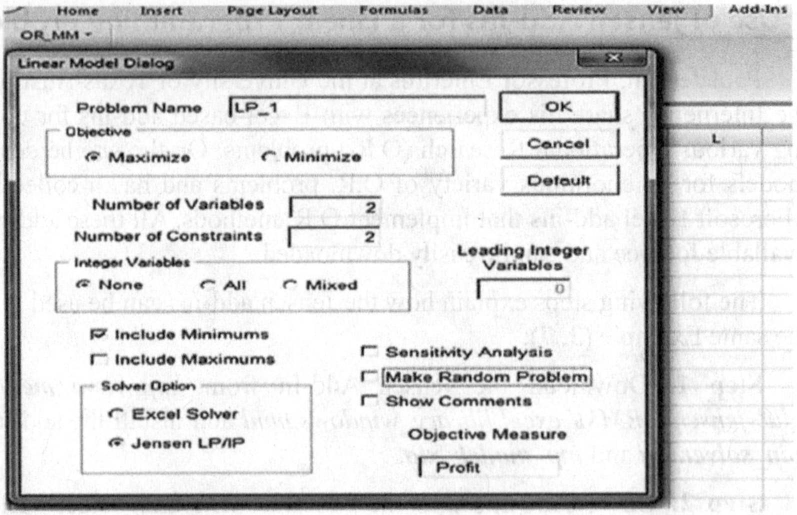

Fig. 3.9

Step 4. Enter all the necessary information as follows:

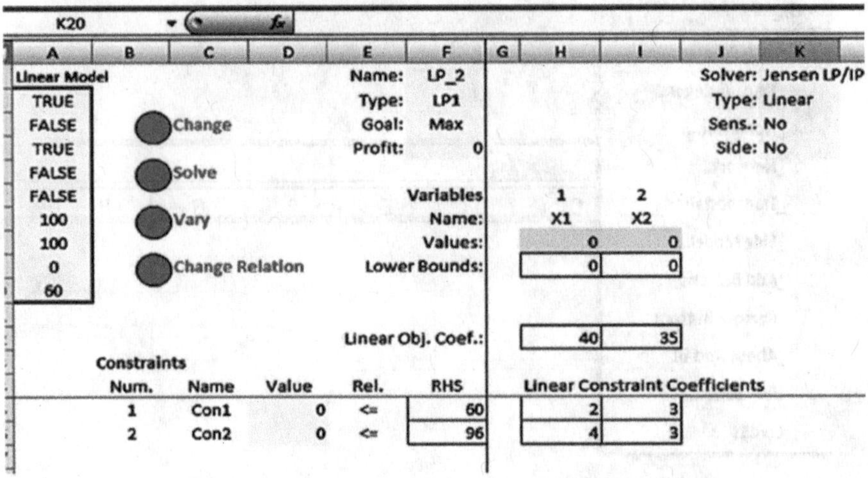

	A	B	C	D	E	F	G	H	I	J	K
	Linear Model				Name:	LP_2				Solver:	Jensen LP/IP
	TRUE				Type:	LP1				Type:	Linear
	FALSE		Change		Goal:	Max				Sens.:	No
	TRUE				Profit:	0				Side:	No
	FALSE		Solve								
	FALSE					Variables		1	2		
	100		Vary			Name:		X1	X2		
	100					Values:		0	0		
	0		Change Relation		Lower Bounds:			0	0		
	60										
					Linear Obj. Coef.:			40	35		
		Constraints									
		Num.	Name	Value	Rel.	RHS	Linear Constraint Coefficients				
		1	Con1	0	<=	60		2	3		
		2	Con2	0	<=	96		4	3		

Fig. 3.10

Step 5. And click solve.

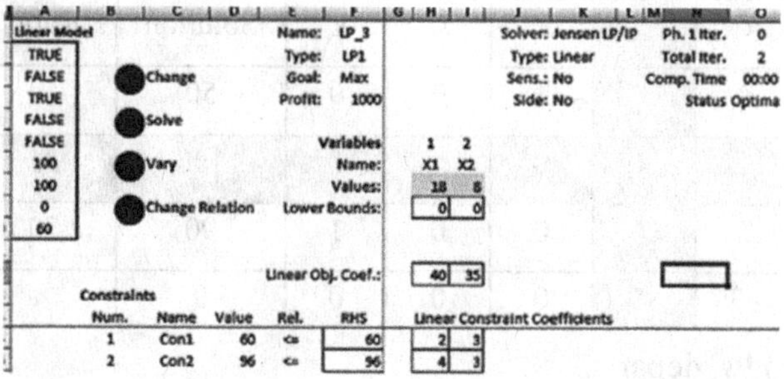

Fig. 3.11

It can be clearly seen that the problem has an optimal solution and the number of iterations required in reaching this solution is 2.

Example 3.8. *Max z = 3x₁ + 9x₂*

Subject to the constraints

$$2x_1 + x_2 \leq 50$$
$$x_1 + 4x_2 \leq 100$$
$$2x_1 + 3x_2 \leq 90$$
$$x_1, x_2 \geq 0$$

Solution: The standard form to the given problem is

Max $z = 3x_1 + 9x_2$

Subject to the constraints

$$2x_1 + x_2 + s_1 = 50$$
$$x_1 + 4x_2 + s_2 = 100 \qquad \qquad ...(3.37)$$
$$2x_1 + 3x_2 + s_3 = 90$$
$$x_1, x_2, s_1, s_2, s_3 \geq 0$$

(3.37) can also be rewritten as

$$2x_1 + x_2 + s_1 = 50$$
$$x_1 + 4x_2 + s_2 = 100$$
$$2x_1 + 3x_2 + s_3 = 90$$
$$z - 3x_1 - 9x_2 = 0$$
$$x_1, x_2, s_1, s_2, s_3 \geq 0$$

Initial Table

Basic	x_1	x_2	s_1	s_2	s_3	Solution	Ratio
s_1	2	1	1	0	0	50	50
s_2	1	4	0	1	0	100	25
s_3	2	3	0	0	1	90	30
z	−3	−9	0	0	0	0	

x_2 enters and s_2 departs.

First Iteration

Basic	x_1	x_2	s_1	s_2	s_3	Solution	Ratio
s_1	1.75	0	1	−0.25	0	25	14.28571
x_2	0.25	1	0	0.25	0	25	100
s_3	1.25	0	0	−0.75	1	15	12
z	−0.75	0	0	2.25	0	225	

x_1 enters and s_3 departs.

Second and Final Iteration

Basic	x_1	x_2	s_1	s_2	s_3	Solution
s_1	0	0	1	0.8	−1.4	4
x_2	0	1	0	0.4	−0.2	22
x_1	1	0	0	−0.6	0.8	12
z	0	0	0	1.8	0.6	234

Since all the elements in the z row are ≥ 0, the optimality condition is now satisfied. Hence, the optimum solution of the problem is $x_1 = 12$ and $x_2 = 22$, and the corresponding maximum value of z is **234**.

Note: Here all the algorithms have been explained by assuming a maximization problem as a standard problem. If any linear programming problem is of a minimization type, then convert that problem into a maximization type by simply multiplying the objective function coefficients by a negative sign and follow the same procedure as is explained for the maximization problem.

   The following are some important points which are to be noted for simplex calculations:

- *Simplex calculations* always *start with a set of initial basic variables whose coefficient matrix forms an identity matrix.*

- *The solution column in the simplex table, at any stage of the calculation, should always be greater than or equal to zero. This is called the* **feasibility condition** *of the problem.*

- *The prices assigned to slack and surplus variables in the objective function will always be zero, because these variables are used only to convert inequalities into equations; otherwise, they do not play any role in optimizing the objective function.*

- *At any stage of the calculation, the columns corresponding to the basic variables should always provide an identity matrix.*

- *The entries in the z-row for the basic variables should be equal to zero at any stage of the calculation.*

- *The simplex method is an iterative method, and it terminates after a finite number of steps with one of the following conclusions:*

   (*a*) the problem has no feasible solution (this will be explained in the next chapter);

   (*b*) the problem has a finite optimal solution; or

   (*c*) the problem has an unbounded solution (this will be explained in the next chapter).

Exercises

1. What is the role of a slack and surplus variable in a linear programming problem?

2. What is a standard and canonical form of a given linear programming problem?

3. Taking x_1 and x_2 as the basic variables, write the canonical form for the following system of linear equations:

$$4x_1 + x_2 - 6x_3 = 40$$

$$x_1 - x_3 = 8$$

Also, find the value of x_1 and x_2 from the resulting system.

4. Put the following problem into standard form:

(i) Max $z = x_1 + x_2 + x_3$

Subject to

$$x_1 + 4x_2 - x_3 = 20$$
$$2x_2 + 3x_3 \leq 35$$
$$x_1, x_2, x_3 \geq 0$$

(ii) Max $z = x_1 - 3x_2 + x_3$

Subject to

$$-3x_1 + 4x_2 + x_3 \leq 120$$
$$2x_1 + 3x_2 + x_3 \geq 350$$
$$x_1, x_2 \geq 0 \text{ and } x_3 \text{ unrestricted in sign.}$$

(iii) Min $z = x_1 + x_2 - 2x_4$

Subject to

$$x_1 + 4x_2 + x_3 + x_4 = 55$$
$$3x_2 + 4x_4 + x_5 \geq 35$$
$$x_2, x_3, x_4, x_5 \geq 0 \text{ and } x_1 \text{ unrestricted in sign,}$$

(iv) Max $z = 6x_1 + 2x_2 + x_3 + 20$

Subject to

$$4x_1 - 3x_2 - x_3 \leq 100$$
$$x_1 + x_2 + x_3 \leq 150$$
$$10 \leq x_3 \leq 30$$
$$x_1, x_2 \geq 0$$

5. Put the (i) and (iii) parts of problem 4 into canonical form.

6. Explain the steps involved in simplex calculation.

7. What is "pivoting" in simplex calculation?

8. Use the algebraic approach of simplex calculation to solve problem 4(i).

9. Graphically explain the steps of the simplex method for the following problem:

$$\text{Max } z = 2x_1 + 3x_2$$

Subject to

$$6x_1 + 4x_2 \le 48$$
$$2x_1 + 3x_2 \le 30$$
$$x_1 \le 6, x_2 \le 5 \text{ and } x_1, x_2 \ge 0$$

10. Consider the linear program:

$$\text{Max } z = 6x_1 + 5x_3 - x_5$$

Subject to:

$$2x_1 + x_2 + 4x_3 - x_5 = 20$$
$$3x_1 + x_3 + x_4 + 2x_5 = 40$$
$$x_1 + 1.5x_3 + x_5 + x_6 = 35$$
$$x_j \ge 0, j = 1, 2, \ldots, 6$$

(*i*)   Find an initial basic feasible solution and its values.

(*ii*) Convert the given system of equations to the canonical form for carrying out the simplex routine.

(*iii*) Is the current initial basic feasible solution optimal? If not, why?

(*iv*) How would you select the entering variable if the current basic feasible solution is not optimal?

(*v*)   Having chosen an entering variable, how do you select the leaving variable?

(*vi*) What will happen if none of the current basic variables is eligible for leaving the basis?

(*vii*) What is the rule for ensuring that the new basic solution is still feasible?

(*viii*) Without performing the pivot operation, find the values of the new basic variables and check for optimality. If the current basic variables do not provide us an optimal solution, complete the operation and find the optimal basic feasible solution to the given problem.

11. Solve the following linear programming problems by the simplex method:

(*i*) Maximize $Z = 5x_1 + 4x_2 + 7x_3$

Subject to the constraints $x_1 + 2x_2 + x_3 \leq 300$

$$2x_1 + x_2 + 2x_3 \leq 550$$

$$x_1 + 3x_2 + x_3 \leq 500, \quad x_1, x_2, x_3 \geq 0$$

(*ii*) Maximize $Z = 1x_1 - 2x_2 + x_3$

Subject to the constraints $-2x_1 + x_2 + x_3 \leq 5$

$$-1x_1 + x_2 - 2x_3 \leq 8$$

x_1 unrestricted in sign, and $x_2, x_3 \geq 0$.

(*iii*) Maximize $Z = 7x_1 + 5x_2 + 3x_3 + 2x_4$

Subject to the constraints $x_1 + 3x_2 + 2x_3 + x_4 \leq 50$

$$3x_1 + x_2 + 1x_3 + x_4 \leq 60$$

$$x_1, x_2, x_3, x_4 \geq 0$$

(*iv*) Maximize $Z = 2x_1 + 7x_2$

Subject to the constraints $x_1 + 3x_2 \leq 20$

$$1x_1 + 2x_2 \leq 12$$

$$3x_2 \leq 5, \quad x_1, x_2 \geq 0$$

CHAPTER 4

SIMPLEX METHOD-II

4.1. Introduction

As already mentioned in the previous chapter, the simplex algorithm starts with an initial basic feasible solution where the coefficient matrix corresponding to the basic variables forms an identity matrix, but there may be a situation where we will not be able to get such an identity matrix to start the simplex computation. This type of situation mostly occurs when we have constraints with ($\geq$) inequalities or with equalities in the given linear programming problem. In such circumstances, a new dummy variable is introduced in the problem just to provide us an identity matrix for starting the simplex calculation, and such a variable is known as an artificial variable.

Definition: The non-negative variable which is added to the problem so as to have an identity basis matrix to start the simplex calculation is known as an artificial variable (AV). These variables are called artificial because they don't have any meaning for the original problem, and hence the initial basic feasible solution obtained by adding them to the initial problem will not be a basic feasible solution for the original problem; it will only be a basic feasible solution for the new problem. Thus, any solution to the original problem should not contain an artificial variable at a positive level. If the final solution of the problem (when the optimality condition is satisfied and further improvement is not possible) has one or more artificial variables in the basis at the positive level, then no solution exists for the original problem (which may be due to inconsistent constraints or due to the nonexistence of a feasible solution).

The presence of an artificial variable in the problem has to be handled differently. There are two methods for handling the presence of this variable in the problem.

(i) Big M method

(ii) two-phase method

Both methods work in such a way that the artificial variables are driven away from the basis as soon as possible so that we can get a basic feasible solution for the original problem, and hence by further application of simplex method, we will be able to obtain the optimal solution to the original problem. Once artificial variables vanish from the basis, they will never enter again into the basis at a positive level, since the simplex method works in such a way that it always improves the solution in further iterations.

Remark: An artificial variable will not have any effect on the final value of the objective function.

4.2. Big M Method (Penalty Method)

The Big M method is a version of the simplex algorithm that first finds a basic feasible solution by adding "artificial" variables to the problem. The following steps are involved in this method.

(i) Make sure that the RHS of each constraint is non-negative. If required, a constraint may be multiplied by −1 to convert its RHS into positive. Don't forget to reverse the direction of an inequality if it is multiplied by −1.

(ii) Convert the problem into its standard form by adding a suitable variable (s) to each of the inequality constraints.

(iii) An artificial variable $a_i(\geq 0)$ will also be added to the constraints identified as $\geq$ or $=$ at the end of step 1.

(iv) If the given linear programming problem (LPP) is a maximization problem, add (for each artificial variable) a penalty of −M times the artificial variable to the objective function. If the LPP is a minimization problem, then add (for each artificial variable) M times the artificial variable to the objective function, where M denotes a very large positive number. The assignment of a large negative price to the artificial variable in the maximization problem and a large positive price in the minimization problem makes it unfavorable for the artificial variable to enter again into the basis (after being removed from the basis), since simplex method tries to improve the solution in all further iterations and the entry of the variable with a large negative price will deteriorate the solution further.

(v) Express the problem into its canonical form to start the simplex calculation and apply the simplex method in the usual way as was done earlier.

(vi) Once an artificial variable departs from the basis, the column corresponding to that artificial variable can be dropped from the remaining calculations.

(vii) At the end, if *(i)* no artificial variable exists in the basis and the optimality condition is satisfied, then constraints are consistent, there is no redundancy and optimal basic feasible solution exists for the original problem; *(ii)* one or more artificial variables exist on the basis at the positive level and the optimality condition is satisfied, then there is no feasible solution for the original problem; *(iii)* one or more artificial variables exist in the basis at a zero level and the optimality condition is satisfied, then constraints are consistent but redundancy may exist in the constraint, and there is an optimal basic feasible solution to the original problem.

Remark: This method is also referred as the penalty method. The method was first suggested by Charnes and hence is also called the M-Charnes method.

Example 4.1. *An oil company producing a petroleum product requires an input of crude oil A and crude oil B. Each barrel of the final product must contain at least 100 gallons. In this final mix, at least 60 gallons must be of crude oil A and, at most, 50 gallons can be from B. Crude oil B costs $26.13 per gallon, and crude oil A costs $25.50 per gallon. Use simplex method to find out how many gallons of A and B should be used in each barrel of the petroleum product in order to meet the specifications and at the same time minimize the cost.*

Solution: Let x_1 and x_2 be the number of gallons of crude oil A and B, respectively, in a barrel of final product, and then the problem can be formulated as follows:

$$Min \ z = 2550x_1 + 2613x_2$$

Subject to

$$x_1 + x_2 \geq 100$$
$$x_1 \geq 60$$
$$x_2 \leq 50$$
$$x_1, x_2 \geq 0 \qquad \qquad ...(4.1)$$

Introducing surplus variables ($s_1, s_2 \geq 0$) and a slack variable ($s_3 \geq 0$) in the problem so that the standard form of the above problem is

$$Min \ z = 2550x_1 + 2613x_2$$

Subject to

$$x_1 + x_2 - s_1 = 100$$
$$x_1 - s_2 = 60$$
$$x_2 + s_3 = 50$$
$$x_1, x_2, s_1, s_2, s_3 \geq 0 \qquad ...(4.2)$$

Now, introducing the artificial variables $a_1 \geq 0$ and $a_2 \geq 0$ in the first and second constraints, respectively, and assigning the high positive price M to both the artificial variables in the objective function, we get:

$$Min \ z = 2550x_1 + 2613x_2 + 0s_1 + 0s_2 + 0s_3 + Ma_1 + Ma_2$$

Subject to

$$x_1 + x_2 - s_1 + a_1 = 100$$
$$x_1 - s_2 + a_2 = 60$$
$$x_2 + s_3 = 50$$
$$x_1, x_2, s_1, s_2, s_3, a_1, a_2 \geq 0 \qquad ...(4.3)$$

Now, our starting set of basic variables is $x_B = (a_1, a_2, s_3)$. Expressing (4.3) into its canonical form we have:

$$x_1 + x_2 - s_1 + a_1 = 100$$
$$x_1 - s_2 + a_2 = 60$$
$$x_2 + s_3 = 50$$
$$z + (2M - 2550)x_1 + (M - 2613)x_2 - Ms_1 - Ms_2 = 160M$$

$$x_1, x_2, s_1, s_2, s_3, a_1, a_2 \geq 0 \qquad ...(4.4)$$

Basic Variables	x_1	x_2	s_1	s_2	s_3	a_1	a_2	solution	Ratio
a_1	1	1	−1	0	0	1	0	100	100
a_2	1	0	0	−1	0	0	1	60	60
s_3	0	1	0	0	1	0	0	50	
z	2M -2550	M−2613	−M	−M	0	0	0	160 M	

Since all the elements in the z row are not ≤ 0, the optimality condition is yet to be fulfilled. The z coefficient for x_1 is most positive, and it will be the entering variable in the basis, and a_2 corresponding to the minimum ratio will depart from the basis. Clearly x_1 replaces a_2 in the basis in the next table. (The column corresponding to a_2 can be dropped from further calculations, to simplify the calculations, since artificial variables are the only variables which once they depart from the basis will never enter into the basis again.)

Basic Variables	x_1	x_2	s_1	s_2	s_3	a_1	solution	Ratio
a_1	0	1	−1	1	0	1	40	40
x_1	1	0	0	−1	0	0	60	
s_3	0	1	0	0	1	0	50	50
z	0	M−2613	− M	M−2550	0	0	40M + 153000	

Since all the elements in the z row are not ≤ 0, a_1 is replaced by s_2 in the next table. (The column corresponding to a_1 can be dropped from further calculations.)

Basic Variables	x_1	x_2	s_1	s_2	s_3	solution
s_2	0	1	−1	1	0	40
x_1	1	1	−1	0	0	100
s_3	0	1	0	0	1	50
z	0	− 63	− 2550	0	0	255000

Now that all the elements in the z row are ≤ 0, the optimality condition is satisfied. Hence, $x_1 = 100$ and $x_2 = 0$ is the optimum solution of the problem, and the corresponding value of z is $2550.00. So, there will be 100 gallons of crude A and zero gallons of crude B in a barrel of final product with a total cost of $2550.00.

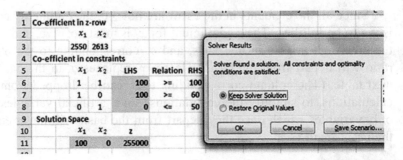

Fig. 4.1

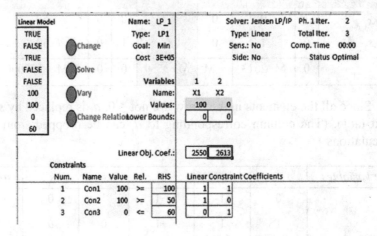

Fig. 4.2

Example 4.2. *Use penalty method to solve the following LPP.*

$$Min\ z = 2x_1 + 2x_2 - 5x_3$$

Subject to

$$3x_1 + 2x_2 - 4x_3 = 10$$

$$x_1 - x_2 + 3x_3 = 60$$

$$x_1, x_2, x_3, \geq 0$$

Solution: By introducing the artificial variables $a_1 \geq 0$ and $a_2 \geq 0$ in the first and second constraints, respectively, we have:

$$Min\ z = 2x_1 + 2x_2 - 5x_3 + Ma_1 + Ma_2$$

Subject to

$$3x_1 + 2x_2 - 4x_3 + a_1 = 10$$

$$x_1 - x_2 + 3x_3 + a_2 = 60$$

$x_1, x_2, x_3, a_1, a_2 \geq 0$ and M is a large positive number.   ...(4.5)

Taking a_1 and a_2 as the starting basic variables, the previous problem can be written in canonical form as:

$$3x_1 + 2x_2 - 4x_3 + a_1 = 10$$

$$x_1 - x_2 + 3x_3 + a_2 = 60$$

$$z + (4M - 2)x_1 + (M - 2)x_2 - (M - 5)x_3 = 70M$$

$$x_1, x_2, x_3, a_1, a_2 \geq 0$$ ...(4.6)

Basic Variables	x_1	x_2	x_3	a_1	a_2	solution	Ratio
a_1	3	2	-4	1	0	10	3.333
a_2	1	-1	3	0	1	60	60
z	$4M - 2$	$M - 2$	$-M + 5$	0	0	70 M	

Not all elements in the z row are ≤ 0; a_2 leaves and a_2 enters the basis.

Basic Variables	x_1	x_2	x_3	a_2	solution	Ratio
x_1	1	0.667	-1.333	0	3.333	
a_2	0	-1.667	4.333	1	56.667	13.078
z	0	$-1.666M - 0.666$	$4.333M + 2.334$	0	$56.667M + 6.666$	

Not all elements in the z row are ≤ 0; a_2 leaves and x_3 enters the basis.

Basic Variables	x_1	x_2	x_3	solution	Ratio
x_1	1	0.153	0	20.769	20.769
x_3	0	-0.384	1	13.077	
z	0	0.231	0	-23.846	

Not all elements in the z row are ≤ 0; x_1 leaves and x_2 enters the basis.

Basic Variables	x_1	x_2	x_3	solution
x_2	6.5	1	0	135
x_3	2.5	0	1	65
z	-1.5	0	0	-55

Since all the elements in the z row are ≤ 0, the optimality condition is satisfied. Hence, $x_1 = 0$, $x_2 = 135$, and $x_3 = 65$ is the optimum solution of the problem, and the corresponding value of z is -55.

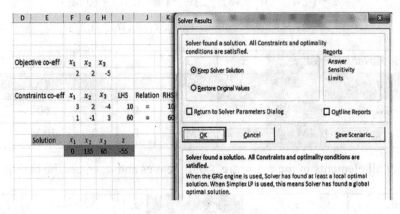

Fig. 4.3

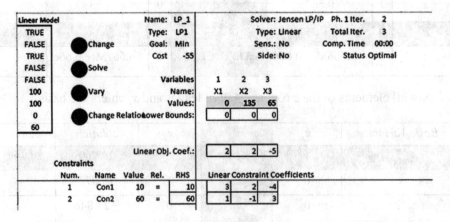

Fig. 4.4

Example 4.3. *Use the Big **M**-Method to solve the following LPP:*

$$Max\ z = 10x_1 + 15x_2$$

Subject to

$$x_1 + 0.75\ x_2 \le 4$$
$$2x_1 + 5\ x_2 \ge 20$$
$$2x_1 + 2\ x_2 = 10$$
$$x_1, x_2, \ge 0$$

Solution: The standard form to the given problem is:

$$Max\ z = 10x_1 + 15x_2$$

Subject to

$$x_1 + 0.75x_2 + s_1 = 4$$
$$2x_1 + 5x_2 - s_2 = 20$$
$$2x_1 + 2x_2 = 10$$
$$x_1, x_2, s_1, s_2 \ge 0 \qquad\qquad ...(4.7)$$

Introducing the artificial variables a_1 and a_2 in the second and third constraints of (4.7), we have:

$$Max\ z = 10x_1 + 15x_2 - Ma_1 - Ma_2$$

Subject to

$$x_1 + 0.75x_2 + s_1 = 4$$
$$2x_1 + 5x_2 - s_2 + a_1 = 20$$
$$2x_1 + 2x_2 + a_2 = 10$$
$$x_1, x_2, s_1, s_2, a_1, a_2 \ge 0 \qquad\qquad ...(4.8)$$

Rewriting (4.8) into its canonical form, we have:

$$x_1 + 0.75x_2 + s_1 = 4$$
$$2x_1 + 5x_2 - s_2 + a_1 = 20$$
$$2x_1 + 2x_2 + a_2 = 10$$
$$z - (4M + 10)x_1 - (15 + 7M)x_2 + Ms_2 = -30M$$
$$x_1, x_2, s_1, s_2, a_1, a_2 \ge 0 \qquad\qquad ...(4.9)$$

Basic Variables	x_1	x_2	s_1	s_2	a_1	a_2	Solution	Ratio
s_1	1	0.75	1	0	0	0	4	5.333
a_1	2	5	0	−1	1	0	20	4
a_2	2	2	0	0	0	1	10	5
z	− 4M − 10	− 7M − 15	0	M	0	0	− 30M	

Basic Variables	x_1	x_2	s_1	s_2	a_2	Solution	Ratio
s_1	0.7	0	1	0.15	0	1	1.429
x_2	0.4	1	0	− 0.2	0	4	10
a_2	1.2	0	0	0.4	1	2	1.667
z	− 1.2M − 4	0	0	− 0.4M − 3	0	− 2M + 60	

Basic Variables	x_1	x_2	s_1	s_2	a_2	Solution	Ratio
x_1	1	0	1.429	0.214	0	1.429	6.667
x_2	0	1	− 0.571	− 0.285	0	3.429	
a_2	0	0	− 1.714	0.1428	1	0.286	2
z	0	0	1.715M + 5.716	− 0.143M − 2.144	0	− 0.285M + 65.716	

Basic Variables	x_1	x_2	s_1	s_2	Solution	Ratio
x_1	1	0	4	0	1	0.25
x_2	0	1	− 4	0	4	
s_2	0	0	− 12	1	2	
z	0	0	− 20	0	70	

Basic Variables	x_1	x_2	s_1	s_2	Solution
s_1	0.25	0	1	0	0.25
x_2	1	1	0	0	5
s_2	3	0	0	1	5
z	5	0	0	0	75

Since all the elements in the z row are ≥ 0, the optimality condition is satisfied. Hence, $x_1 = 0$ and $x_2 = 5$ is the optimum solution of the problem, and the corresponding value of z is 75.

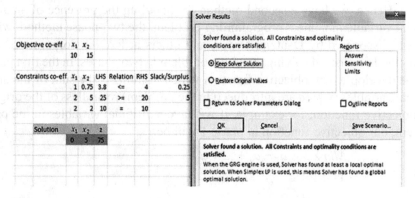

Fig. 4.5

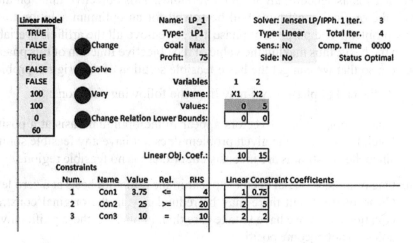

Fig. 4.6

4.3. Two-Phase Method

In the Big M method, by assigning a very large negative price to the artificial variable in the maximization problem, we try to make the artificial variable unprofitable to be there in the basis and intend to drive it out of the basis to get the basic feasible solution for the original problem. But if this problem is solved on the computer, then a very large numerical value has to be assigned to M, which sometimes can create problems and result in an incorrect answer.

Hence, we have a second method for handling the presence of an artificial variable in a linear programming problem, the two-phase method, which was discovered by Dantzig, Orden, and others at RAND Corporation. This method consists of solving the given problem in two phases. In the first phase, we develop a new objective function which minimizes the sum of the artificial variables (which are introduced in the problem) regardless of the original objective function; for example, if we have n artificial variables in the problem, our objective function in phase 1 will be:

$$Min \ Z^{p1} = a_1 + a_2 + ... + a_n$$

or $\quad Max \ Z^{p1} = - a_1 - a_2 - ... - a_n$

Here a price of 1 (in a minimization problem) or -1 (in a maximization problem) is assigned to all artificial variables. This objective function along with the original constraints will be solved for an optimum solution to this new problem. The purpose of phase 1 is to remove all the artificial variables from the basis, thus making the value of the objective function of this phase to be zero so that we can get the basic feasible solution of the original problem.

At the end of phase 1, we may have the following situations:

(i) One or more artificial vectors appear in the optimum basis at a positive level. In this case, the given problem does not have any feasible solution since the constraints are inconsistent or there is no feasible region.

(ii) One or more artificial vectors appear in the optimum basis at a zero level. Go to phase 2, but there may be redundancy in the original constraint equations. Here we have to take care that, in phase 2, these artificial variables do not become positive.

(iii) No artificial vector appears in the optimum basis. In this case also proceed to phase 2. Here constraints are consistent, and there is no redundancy in the constraints.

The phase 2 calculation starts with the objective function given in the original problem by assigning zero prices to the artificial variables which still exist in the basis at the zero level. The basic feasible solution appearing in the final table of phase 1 will be considered as the initial basic feasible solution of phase 2; that is the final table of phase 1 will be taken as the first table of phase 2 with different values of z coefficients (since the objective function coefficients have been changed). The artificial variables which are non-basic a t the end of phase 1 are removed from further consideration in phase 2.

Now the simplex method will be applied as usual to get the optimal solution to the original problem by paying special attention to the artificial variables which are appearing at a zero level in the basis of phase 2 (if any). There is a need to make sure that these artificial variables always remain at zero and never become positive in any iteration. To keep them at zero, sometimes we have to change the criterion for deciding about the leaving variable; that is, in the pivot column, if $a_{ik} < 0$, where a_{ik} is the coefficient corresponding to the artificial variable i in pivot column k, then instead of removing the vector according to the usual criterion, we will remove this artificial variable from the basis; otherwise, it will become positive in the next iteration.

Example 4.4. *Use two-phase method to solve Example 4.1.*

Solution: Let x_1 and x_2 be the number of gallons of crude oil A and B in a barrel of final product, and then the problem is formulated as follows:

$$Min \; z = 2550x_1 + 2613x_2$$

Subject to

$$x_1 + x_2 \geq 100$$

$$x_1 \geq 60$$

$$x_2 \leq 50$$

$$x_1, x_2 \geq 0 \qquad \qquad ...(4.10)$$

Introducing the surplus (s_1, s_2) and slack (s_1) variables in the given problem to convert it into the standard form such that the problem becomes:

$$Min \; z = 2550x_2 + 2613x_2$$

Subject to

$$x_1 + x_2 - s_1 = 100$$

$$x_1 - s_2 = 60$$

$$x_2 + s_3 = 50$$

$$x_1, x_2, s_1, s_2, s_3 \geq 0 \qquad \qquad ...(4.11)$$

Consider the following LP problem derived from the original one by adding the artificial variables a_1 and a_2 in the first and second constraints, respectively, to get the initial basic feasible solution and introduce a new objective function which consists of minimizing the sum of the artificial variables.

Phase 1

$$Min\ w = a_1 + a_2$$

Subject to

$$x_1 + x_2 - s_1 + a_1 = 100$$
$$x_1 - s_2 + a_2 = 60$$
$$x_2 + s_3 = 50$$
$$x_1, x_2, s_1, s_2, a_1, a_2 \geq 0 \qquad \qquad ...(4.12)$$

Phase 1 of the problem has an initial basic feasible solution with basic variables being a_1, a_2, and S_3. If the minimum value of $a_1 + a_2$, is 0, then the solution of phase 1 (even if the optimality condition is not satisfied) can be considered as the initial basic feasible solution of the original problem. If the minimum value of $a_1 + a_2$ is greater than 0, then the original problem does not have any feasible solution.

Now, we construct tables to solve the phase 1 problem. The objective value w. should be written in terms of non-basic variables to convert it into canonical form.

$$x_1 + x_2 - s_1 + a_1 = 100$$
$$x_1 - s_2 + a_2 = 60$$
$$x_2 + s_3 = 50$$
$$w + 2x_1 + x_2 - s_1 - s_2 = 160$$
$$x_1, x_2, s_1, s_2, s_3, a_1, a_2 \geq 0 \qquad \qquad ...(4.13)$$

Basic Variables	x_1	x_2	s_1	s_2	s_3	a_1	a_2	solution	Ratio
a_1	1	1	−1	0	0	1	0	100	100
a_2	1	0	0	−1	0	0	1	60	60
s_3	0	1	0	0	1	0	0	50	
w	2	1	−1	−1	0	0	0	160	

Basic Variables	x_1	x_2	s_1	s_2	s_3	a_1	a_2	solution	Ratio
a_1	0	1	−1	1	0	1	−1	40	40
x_1	1	0	0	−1	0	0	1	60	
s_3	0	1	0	0	1	0	0	50	50
w	0	1	−1	1	0	0	−2	40	

Basic Variables	x_1	x_2	s_1	s_2	s_3	a_1	a_2	solution
x_2	0	1	−1	1	0	1	−1	40
x_1	1	0	0	−1	0	0	1	60
s_3	0	0	1	−1	1	−1	1	10
w	0	0	0	0	0	−1	−1	0

The optimal value of the phase 1 problem is $w = 0$. So, the original problem is feasible, and the initial basic feasible solution of the phase 2 problem is $x_1 = 60$, $x_2 = 40$, $s_1 = 0$, $s_2 = 0$, and $s_3 = 10$. To start phase 2, we replace the w row in the final table of phase 1 by the original objective function, after dropping the columns of the artificial variables, and we have:

Phase 2

Basic Variables	x_1	x_2	s_1	s_2	s_3	solution	Ratio
x_2	0	1	−1	1	0	40	
x_1	1	0	0	−1	0	60	
s_3	0	0	1	−1	1	10	
z	−2550	−2613	0	0	0	0	

Doing the elementary row operations, the coefficients of the basic variables are eliminated from the z row, and we have:

Basic Variables	x_1	x_2	s_1	s_2	s_3	solution	Ratio
x_2	0	1	−1	1	0	40	40
x_1	1	0	0	−1	0	60	
s_3	0	0	1	−1	1	10	
z	0	0	−2613	63	0	257520	

Basic Variables	x_1	x_2	s_1	s_2	s_3	solution	Ratio
s_2	0	1	−1	1	0	40	
x_1	1	1	−1	0	0	100	
s_3	0	1	0	0	1	50	
z	0	−63	−2550	0	0	255000	

The optimum condition is now satisfied. Therefore, the optimum solution of the problem is $x_1 = 100$ and $x_2 = 0$, and the corresponding value of z is 255,000. This solution is in agreement with the solution obtained by using Big M method.

Example 4.5. *Use the two-phase method to solve Example 4.2.*

Solution: Consider the following LP problem derived from the original one by adding artificial variables a_1 and a_2 to the first and second constraints, respectively. A new objective function which consists of minimizing $(a_1 + a_2)$ is introduced in the problem and we have:

Phase 1

$$Min \ w = a_1 + a_2$$

Subject to

$$3x_1 + 2x_2 - 4x_3 + a_1 = 10$$

$$x_1 - x_2 + 3x_3 + a_2 = 60$$

$$x_1, x_2, x_3, a_1, a_2 \geq 0 \qquad \qquad ...(4.14)$$

Taking a_1 and a_2 as the starting basic variables, the canonical form of (4.14) can be written as:

$$3x_1 + 2x_2 - 4x_3 + a_1 = 10$$

$$x_1 - x_2 + 3x_3 + a_2 = 60$$

$$w + 4x_1 + x_2 - x_3 = 70$$

$$x_1, x_2, x_3, a_1, a_2 \geq 0 \qquad \qquad ...(4.15)$$

Basic Variables	x_1	x_2	x_3	a_1	a_2	solution	Ratio
a_1	3	2	−4	1	0	10	3.333333
a_2	1	−1	3	0	1	60	60
w	4	1	−1	0	0	70	

Basic Variables	x_1	x_2	x_3	a_1	a_2	solution	Ratio
x_1	1	0.666667	−1.33333	0.333333	0	3.333333	
a_2	0	−1.66667	4.333333	−0.33333	1	56.66667	13.0769
w	0	−1.66667	4.333333	−1.33333	0	56.66667	

Basic Variables	x_1	x_2	x_3	a_1	a_2	solution
x_1	1	0.153846	0	0.230769	0.307692	20.76923
x_3	0	−0.38462	1	−0.07692	0.230769	13.07692
w	0	0	0	−1	−1	0

The optimal value of the phase 1 problem is $w = 0$. So, the original problem is feasible, and the initial basic feasible solution of the phase 2 problem is $x_1 = 20.76923$, $x_2 = 0$, $x_3 = 13.07692$. To start phase 2, we use the original objective function and, after dropping the columns of the artificial variables from the final table of phase 1, consider it as the initial tableau of phase 2, that is:

Phase 2

Basic Variables	x_1	x_2	x_3	solution	Ratio
x_1	1	0.153846	0	20.76923	
x_3	0	−0.38462	1	13.07692	
z	−2	−2	5	0	

After doing the elementary row operations, the coefficients of the basic variables are eliminated from the z row and hence we get:

Basic Variables	x_1	x_2	x_3	solution
x_1	1	0.153846	0	20.76923
x_3	0	− 0.38462	1	13.07692
z	0	0.230792	0	− 23.8461

Basic Variables	x_1	x_2	x_3	solution
x_2	6.5	1	0	135
x_3	2.5	0	1	65
z	−1.5	0	0	−55

The optimality condition is now satisfied. Therefore, the optimum solution of the problem is $x_1 = 0$, $x_2 = 135$, and $x_3 = 65$, and the corresponding value of z is −55. This solution is also in agreement with the solution obtained by using the Big M method.

Example 4.6. *Solve Example 4.3 using the two-phase method.*

Solution: The standard form to the given problem is:

$$Mix\ z = 10x_1 + 15x_2$$

Subject to

$$x_1 + 0.75x_2 + s_1 = 4$$
$$2x_1 + 5x_2 - s_2 = 20$$
$$2x_1 + 2x_2 = 10$$
$$x_1, x_2, s_1, s_2 \geq 0 \qquad \qquad ...(4.16)$$

Consider the following LP problem derived from the original one by introducing artificial variables in the second and the third constraints, with a new objective function which consists of minimizing the sum of the artificial variables.

Phase 1

$$Min\ w = a_1 + a_2$$

Subject to

$$x_1 + 0.75x_2 + s_1 = 4$$
$$2x_1 + 5x_2 - s_2 + a_1 = 20$$
$$2x_1 + 5x_2 + a_2 = 10$$
$$x_1, x_2, s_1, s_2, a_1, a_2 \geq 0 \qquad \qquad ...(4.17)$$

We start with an initial basic feasible solution with basic variables being s_1, a_1, and a_2. If the minimum value of $a_1 + a_2$ is 0, then the solution of phase 1 (even if it is not optimal) will be the initial basic feasible solution of the original problem. If the minimum value of $a_1 + a_2$ is greater than 0, then the original problem is not feasible.

Now we construct tables to solve the phase 1 problem. The objective value w. should be written in terms of non-basic variables to convert it into canonical form.

$$x_1 + 0.75x_2 + s_1 = 4$$
$$2x_1 + 5x_2 - s_2 + a_1 = 20$$
$$2x_1 + 2x_2 + a_2 = 10$$
$$w + 4x_1 + 7x_2 - s_2 = 30$$
$$x_1, x_2, s_1, s_2, a_1, a_2 \geq 0.$$

Basic Variables	x_1	x_2	s_1	s_2	a_1	a_2	Solution	Ratio
s_1	1	0.75	1	0	0	0	4	5.333333
a_1	2	5	0	−1	1	0	20	4
a_2	2	2	0	0	0	1	10	5
w	4	7	0	−1	0	0	30	

Basic Variables	x_1	x_2	s_1	s_2	a_1	a_2	Solution	Ratio
s_1	0.7	0	1	0.15	−0.15	0	1	1.428571
x_2	0.4	1	0	− 0.2	0.2	0	4	10
a_2	1.2	0	0	0.4	− 0.4	1	2	1.666667
w	1.2	0	0	0.4	−1.4	0	2	

Basic Variables	x_1	x_2	s_1	s_2	a_1	a_2	Solution	Ratio
x_1	1	0	1.428571	0.214286	−0.21429	0	1.428571	6.666667
x_2	0	1	− 0.57143	−0.28571	0.285714	0	3.428571	
a_2	0	0	−1.71429	0.142857	− 0.14286	1	0.285714	2
w	0	0	−1.71429	0.142857	−1.14286	0	0.285714	

Basic Variables	x_1	x_2	s_1	s_2	a_1	a_2	Solution
x_1	1	0	4	0	0	−1.5	1
x_2	0	1	−4	0	0	2	4
s_2	0	0	−12	1	−1	7	2
w	0	0	0	0	−1	−1	0

The optimal value of the phase 1 problem is $w = 0$. So, the original problem is feasible, and the initial basic feasible solution of the phase 2 problem is $x_1 = 1$, $x_2 = 4$. To start phase 2, we use the original objective function, and after dropping the columns of the artificial variables from the final table of phase 1, the initial tableau of phase 2 is as follows:

Phase 2

Basic Variables	x_1	x_2	s_1	s_2	Solution
x_1	1	0	4	0	1
x_2	0	1	−4	0	4
s_2	0	0	−12	1	2
z	−10	−15	0	0	0

Doing the elementary row operations, the coefficients of the basic variables are eliminated from the z row, and we have:

Basic Variables	x_1	x_2	s_1	s_2	Solution	Ratio
x_1	1	0	4	0	1	0.25
x_2	0	1	−4	0	4	
s_2	0	0	−12	1	2	
z	0	0	−20	0	70	

Basic Variables	x_1	x_2	s_1	s_2	Solution
s_1	0.25	0	1	0	0.25
x_2	1	1	0	0	5
s_2	3	0	0	1	5
z	5	0	0	0	75

The optimum condition is now satisfied. Therefore, the optimum solution of the problem is $x_1 = 0$ *and* $x_2 = 5$, and the corresponding value maximum value of z is 75. This solution is also in agreement with the solution obtained by using the Big M method.

4.4. Degeneracy in Linear Programming Problems

Degeneracy in a linear programming problem occurs if in any given basic feasible solution, at least one of the basic variables takes on a zero value. This is caused by redundant constraint(s) and could cost the simplex method extra iterations. Degeneracy may occur at any stage of the simplex calculation. An initial basic feasible solution will be degenerate if the RHS of any constraint is zero. If the minimum ratio for determining the leaving variable at any stage of the simplex calculation occurs at more than one place, then the next table will also have a degenerate basic feasible solution. If in the final table at least one of the basic variables vanishes, then the problem will be said to have a degenerate optimal basic feasible solution.

If no degeneracy occurs in any iteration of the simplex method, then the optimal solution of the problem can be found in a finite number of steps by moving from one basis to another and changing a single vector in the basis each time to get the new one. Since there are a finite number of bases and since the simplex method improves the solution in every step, no basis is ever repeated and ultimately the optimal solution is reached in a finite number of iterations or we get an indication that problem has an unbounded solution.

But sometimes when degeneracy is present in the problem, we get entered into cycling, that is, the same set of bases keep repeating when we move from one to another without changing or improving the value of the objective function at any iteration and we get stuck in an infinite loop. In this situation we never get an optimal solution and need to handle the problem properly so that no basis is ever repeated and cycling does not occur. Thus, we discuss here the change in the computation procedure required to avoid cycling when degeneracy is present in the solution. This is better explained by the following examples:

Example 4.7.   *Maz* $z = 10\,x_1 + 15x_2$

Subject to

$$2x_1 + 3x_2 \leq 60$$
$$0.5x_1 + x_2 \leq 30$$
$$x_2 \leq 20$$
$$x_1, x_2 \geq 0$$

Solution: Converting the given problem into its standard form, we have:

$$Maz \; z = 10 \, x_1 + 15x_2$$

Subject to

$$2x_1 + 3x_2 + s_1 = 60$$
$$0.5x_1 + x_2 + s_2 = 30$$
$$x_2 + s_3 = 20$$
$$x_1, x_2, s_1, s_2, s_3 \geq 0$$

Taking s_1 and s_2 as the initial basic variables and writing the z equation in terms of the non-basic variables x_1 and x_2, we have:

$$2x_1 + 3x_2 + s_1 = 60$$
$$0.5x_1 + x_2 + s_2 = 30$$
$$x_2 + s_3 = 20$$
$$z - 10 \, x_1 - 15x_2 = 0$$
$$x_1, x_2, s_1, s_2, s_3 \geq 0$$

Basic Variables	x_1	x_2	s_1	s_2	s_3	Solution	Ratio
s_1	2	3	1	0	0	60	20
s_2	0.5	1	0	1	0	30	30
s_3	0	1	0	0	1	20	20
z	−10	−15	0	0	0	0	

There is a tie in the choice of the leaving variable. Either s_1 or s_3 can leave the basis since the minimum ratio is the same for both of them. Here choice can be made arbitrarily or the smallest i^{th} index can be considered. It is to be noted that when a tie occurs and we select one out of two as the leaving variable, then the other variable in the next iteration become zero, that is, the solution becomes degenerate in the next iteration.

Option I: Randomly allowing s_1 to leave the basis, we have:

Basic Variables	x_1	x_2	s_1	s_2	s_3	Solution
x_2	0.666667	1	0.333333	0	0	60
s_2	− 0.16667	0	− 0.333333	1	0	10
s_3	− 0.66667	0	− 0.333333	0	1	0
z	0	0	5	0	0	300

Since all the values of z are ≥ 0, the optimality condition is now satisfied and the optimum solution of the problem is $x_1 = 0$ *and* $x_2 = 20$, and the corresponding value of z is 300.

Option II: Selecting s_3 to leave the basis instead of s_1, we have:

Basic Variables	x_1	x_2	s_1	s_2	s_3	Solution	Ratio
s_1	2	0	1	0	−3	0	0
s_2	0.5	0	0	1	−1	10	20
x_2	0	1	0	0	1	20	
z	−10	0	0	0	15	300	

Basic Variables	x_1	x_2	s_1	s_2	s_3	Solution
x_1	1	0	0.5	0	−1.5	0
s_2	0	0	− 0.25	1	− 0.25	10
x_2	0	1	0	0	1	20
z	0	0	5	0	0	300

Since all the values of z are ≥ 0, the optimality condition is now satisfied and the optimum solution of the problem is $x_1 = 0$, $x_2 = 20$, and the corresponding value of z is 300. The solution is in agreement with the first option, but an extra iteration is required in this option to come to the final solution; this happens because of the degeneracy in the problem.

Example 4.8. $\qquad$ $Maz\ z = 2.25x_1 - 18x_2 - 60x_3 + 0.5x_4$

Subject to

$$0.25x_1 + 9x_2 - 8x_3 - x_4 \geq 0$$
$$0.5x_1 + 3x_2 - 12x_3 - 0.5x_4 \geq 0$$
$$x_4 \geq 1$$
$$x_1, x_2, x_3, x_4 \geq 0$$

Solution: The standard form of the given linear programming problem is:

$$Maz\ z = 2.25x_1 - 18x_2 - 60x_3 + 0.5x_4$$

Subject to

$$0.25x_1 + 9x_2 - 8x_3 - x_4 + s_1 = 0$$
$$0.5x_1 + 3x_2 - 12x_3 - 0.5x_4 + s_2 = 0$$
$$x_4 + s_3 = 1$$
$$x_1, x_2, x_3, x_4, s_1, s_2, s_3 \geq 0$$

Taking $s_1, s_2,$ and s_3 as the initial basic variables and writing the z equation in terms of the non-basic variables $x_1, x_2, x_2,$ and x_4 we have:

$$0.25x_1 + 9x_2 - 8x_3 - x_4 + s_1 = 0$$
$$0.5x_1 + 3x_2 - 12x_3 - 0.5x_4 + s_2 = 20$$
$$x_4 + s_3 = 1$$
$$z - 2.25x_1 + 18x_2 + 60x_3 - 0.5x_4 = 0$$
$$x_1, x_2, x_3, x_4, s_1, s_2, s_3 \geq 0$$

Basic Variables	x_1	x_2	x_3	x_4	s_1	s_2	s_3	Solution	Ratio
s_1	0.25	9	−8	−1	1	0	0	0	0
s_2	0.5	3	−12	−0.5	0	1	0	0	0
s_3	0	0	0	1	0	0	1	1	
z	−2.25	18	60	−0.5	0	0	0	0	

Basic Variables	x_1	x_2	x_3	x_4	s_1	s_2	s_3	Solution	Ratio
x_1	1	36	−32	−4	4	0	0	0	
s_2	0	−15	4	1.5	−2	1	0	0	0
s_3	0	0	0	1	0	0	1	1	
z	0	99	−12	−9.5	9	0	0	0	

Basic Variables	x_1	x_2	x_3	x_4	s_1	s_2	s_3	Solution	Ratio
x_1	1	−84	0	8	−12	8	0	0	0
x_3	0	−3.75	1	0.375	−0.5	0.25	0	0	0
s_3	0	0	0	1	0	0	1	1	1
z	0	54	0	−5	3	3	0	0	

Basic Variables	x_1	x_2	x_3	x_4	s_1	s_2	s_3	Solution	Ratio
x_4	0.125	−10.5	0	1	−1.5	1	0	0	0
x_3	−0.0469	0.188	1	0	0.063	−0.125	0	0	0
s_3	−0.125	10.5	0	0	1.5	−1	1	1	0.667
z	0.625	1.5	0	0	−4.5	8	0	0	

Basic Variables	x_1	x_2	x_3	x_4	s_1	s_2	s_3	Solution	Ratio
x_4	−1	−6	24	1	0	−2	0	0	
s_1	−0.75	3	16	0	1	−2	0	0	
s_3	1	6	−24	0	0	2	1	1	1
z	−2.75	15	72	0	0	−1	0	0	

Basic Variables	x_1	x_2	x_3	x_4	s_1	s_2	s_3	Solution
x_4	0	0	0	1	0	0	1	1
s_1	0	7.5	−2	0	1	−0.5	0.75	0.75
x_1	1	6	−24	0	0	2	1	1
z	0	31.5	6	0	0	4.5	2.75	2.75

It can be seen from the previous calculations that a number of iterations have been repeated with same value of basic variables and also without improving the value of the objective function before it finally comes to the optimum solution. At the optimal point, $x_1 = 1$, $x_2 = 0$, $x_3 = 0$, $and\ x_4 = 1$, and corresponding maximum value of z is 2.75.

4.4.1. Perturbation Method for the Resolution of Degeneracy Problems in LPPs

This method will tell us which variable is to be taken out of the basis when the minimum ratio for deciding the leaving variable is not unique. This will not only help us in avoiding the formation of a cycle but also reduce the unnecessary repetition of iteration(s). The following are the steps involved in this method:

1. Let the minimum ratio occur with the basic variables s_1 $and\ s_2$. Rearrange the columns of s_1 $and\ s_2$. The priority is first given to the column of slack variables, then artificial variables, and then finally to the decision variables.

2. All the elements of each row are now divided by the corresponding element of that row in the pivot column.

3. The quotients are compared between the tied rows for each column.

4. The comparison is stopped at the column where the quotients are unequal, and the row having an algebraically smaller ratio is selected as the pivot row.

5. After selection of the key row, the normal simplex procedure is resumed.

Example 4.9. *Solving the same problem given in Example 4.8 by using the perturbation method, we have the following table where a tie in leaving variables occurs—*

Basic Variables	x_1	x_2	x_3	x_4	s_1	s_2	s_3	Solution	Ratio
s_1	0.25	9	-8	-1	1	0	0	0	0
s_2	0.5	3	-12	-0.5	0	1	0	0	0
s_3	0	0	0	1	0	0	1	1	
z	-2.25	18	60	-0.5	0	0	0	0	

To decide the leaving variable, rearrange the columns of s_1 and s_2;

Basic Variables	s_1	s_2	s_3	x_1	x_2	x_3	x_4
s_1	1	0	0	0.25	9	-8	-1
s_2	0	1	0	0.5	3	-12	-0.5

Dividing each row by the corresponding element of that row in the pivot column, we have:

Basic Variables	s_1	s_2	s_3	x_1	x_2	x_3	x_4
s_1	4	0	0	1	36	-32	-4
s_2	0	2	0	1	6	-24	-1

Now, we compare the values between these two rows for each column from left to right. It can be seen that in the first column, the second row has a lesser value than the first, and hence s_2 should be our leaving variable.

Basic Variables	x_1	x_2	x_3	x_4	s_1	s_2	s_3	Solution	Ratio
s_1	0	7.5	-2	-0.75	1	-0.5	0	0	
x_1	1	6	-24	1	0	2	0	0	
s_3	0	0	0	1	0	0	1	1	1
z	0	31.5	6	-2.75	0	4.5	0	0	

Basic Variables	x_1	x_2	x_3	x_4	s_1	s_2	s_3	Solution	Ratio
s_1	0	7.5	-2	0	1	-0.5	0.75	0.75	
x_1	1	6	-24	0	0	2	1	1	
x_4	0	0	0	1	0	0	1	1	
z	0	31.5	6	0	0	4.5	2.75	0.275	

This solution is in agreement with the solution obtained by not using the perturbation technique. Here the total number of iterations required is 3 (including the initial table) as compared to 6, which were required to get the optimal solution if this method was not used.

Remarks on the problem of degeneracy in LP problems

1. At any stage of the simplex calculation, if there is more than one choice of variable to leave the basis, then the subsequent table will set one or more of the basic variables equal to zero in the associated basic feasible solution. Such an associated basic feasible solution is called **a degenerate basic feasible solution.**

2. It is possible that a pivot on a degenerate tableau does not change the associated values of the basic feasible solution and its value of z. Such a pivot is called a "degenerate pivot," as in the case of the second option of Example 4.7.

Observation (2) is particularly troublesome, since it goes against the very definition of the simplex algorithm and opens the door to the possibility of an infinite sequence of degenerate pivots, never terminating with optimality.

4.5. Solving a System of Linear Equations Using the Simplex Method

Consider a system of m simultaneous linear equations in n unkowns:

$$Ax = b$$

Where A is a real coefficient matrix of the order mxn, $x \in R^n$ is a vector of the order $n \times 1$ containing the unknown decision variables, and b is an $m \times 1$ order vector of real constants.

To solve such a system of simultaneous linear equations by using simplex method, we will write it into a standard linear programming problem. Add an artificial variable to each equation and create a dummy objective function, which minimizes the sum of the artificial variables. Also, the given unrestricted decision variables need to be converted into non-negative variables by writing them as a difference of two non-negative variables as follows:

$$x = x' - x'', \text{ where } x' \geq 0, x'' \geq 0.$$

Example 4.10. *Solve the following simultaneous linear equations using the simplex method:*

$$8x_1 + 2x_2 = 10$$
$$5x_1 + 4x_2 = 15$$

Solution: Since the range of x_1 *and* x_2 is not given, they will be considerd as unrestricted variables, and hence we can write:

$$x_1 = x'_1 - x''_1$$
$$x_2 = x'_2 - x''_2$$

Where $x'_1, x''_1, x'_2, x''_2 \geq 0.$

Introduce the artificial variables $a_1 \geq 0$ and $a_2 \geq 0$ to the first and second equations, respectively. Also, at the same time a dummy objective function which minimizes the sum of these artificial variables will be created. Now, solving the given system of simultaneous linear equations will be equivalent to solving the following linear programming problem:

$$Min \; z = a_1 + a_2$$

Subject to

$$8(x'_1 - x''_1) + 2(x'_2 - x''_2) + a_1 = 10$$
$$5(x'_1 - x''_1) + 4(x'_2 - x''_2) + a_2 = 25$$
$$x'_1, x''_1, x'_2, x''_2, a_1, a_2 \geq 0$$

Taking a_1 and a_2 as the initial basic variables, the canonical from of the previous LPP is:

$$8(x'_1 - x''_1) + 2(x'_2 - x''_2) + a_1 = 10$$
$$5(x'_1 - x''_1) + 4(x'_2 - x''_2) + a_2 = 25$$
$$z + 13x'_1 - 13x''_1 + 6x'_2 - 6x''_2 = 35$$
$$x'_1, x''_1, x'_2, x''_2, a_1, a_2 \geq 0$$

Basic Variables	x_1'	x_1''	x_2'	x_2''	a_1	a_2	Solution	Ratio
a_1	8	−8	2	−2	1	0	10	1.25
a_2	5	−5	4	−4	0	1	25	5
z	13	−13	6	−6	0	0	35	

Basic Variables	x_1'	x_1''	x_2'	x_2''	a_2	Solution	Ratio
x^1_1	1	−1	0.25	−0.25	0	1.25	5
a_2	0	0	2.75	−2.75	1	18.75	6.81818
z	0	0	2.75	−2.75	0	18.75	

Basic Variables	x_1'	x_1''	x_2'	x_2''	a_2	Solution	Ratio
x_2'	4	−4	1	−1	0	5	
a_2	−11	11	0	0	1	5	0.45454
z	−11	11	0	0	0	5	

Basic Variables	x_1'	x_1''	x_2'	x_2''	Solution
x_2'	0	0	1	−1	6.81818
x_1''	−1	1	0	0	0.45454
z	0	0	0	0	0

Since all the elements in the z row are ≤ 0, the optimality condition is now satisfied and the optimum solution to the given system of equations is:

$$x_1' = 0, x_1'' = 0.45454, x_2' = 6.81818, x_2'' = 0.$$

And hence, $x_1 = -0.45457$ and $x_2 = 6.81818$.

4.6. Solution of a System of Linear Equations by Using Solver

The following are the steps involved for solving a given system of linear equations in Excel Solver. Consider Example 4.10.

(i) Enter the problem as follows.

		x_1	x_2	LHS	Relation	RHS
equation co-efficients		8	2		=	10
		5	4		=	25
	Solution	x_1	x_2			

(ii) Compute the **LHS** value as the **Sumproduct** of the solution value and the corresponding coefficient of the variables in the given equation.

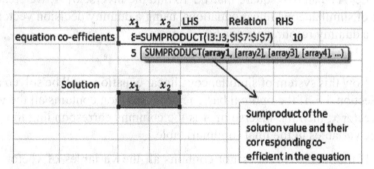

Sumproduct of the solution value and their corresponding co-efficient in the equation

(iii) Now go to Solver, leave the **Set Target Cell** blank, and select the **Value of** cell as shown. Also enter all the other information as done before. Remember to select the simplex option. If the variables are free variables (unrestricted), then don't select the non-negative option.

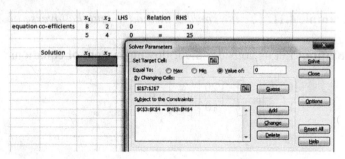

(iv) On clicking solve we have the following output, and this is in agreement with what we had obtained using simplex.

		x_1	x_2	LHS	Relation	RHS
equation co-efficients		8	2	10	=	10
		5	4	25	=	25
	Solution	x_1	x_2			
		-0.45	6.818			

Solver Results

Solver found a solution. All constraints and optimality conditions are satisfied.

Reports

- Keep Solver Solution
- Restore Original Values

Answer
Sensitivity
Limits

OK   Cancel   Save Scenario...   Help

4.7. Inverse of a Matrix Using the Simplex Method

Let **A** be an *nxn* square matrix. To find the inverse of **A**, we create a system of simultaneous linear equations, using a dummy decision vector $x \in R^n$ and a dummy resource vector b.

$$Ax = b$$

Now this system of simultaneous linear equations will be solved as given in Section 4.5. If the optimal basic feasible solution contains all the variables of vector x, then the inverse of A is the columns corresponding to the initial basic variables in the final (optimal) table.

If the final table does not contains all the variables of vector x in the basis we continue with the simplex procedure until all the variables of vector x are in the basis and at the same time the optimality condition is also satisfied, but the solution in the process may or may not be feasible (see Example 4.11).

Example 4.11. *Use the simplex method to find the inverse of* $A = \begin{bmatrix} 10 & 4 \\ 4 & 3 \end{bmatrix}$.

Solution: Let us consider the following system of equations:

$$\begin{bmatrix} 10 & 4 \\ 4 & 3 \end{bmatrix} \begin{bmatrix} x_1 \\ x_2 \end{bmatrix} = \begin{bmatrix} 20 \\ 10 \end{bmatrix}; \begin{bmatrix} x_1 \\ x_2 \end{bmatrix} \geq 0$$

...(4.18)

where $b = \begin{bmatrix} 20 \\ 10 \end{bmatrix}$ is a dummy column vector. Now, introducing the artificial variables $a_1 \geq 0$ and $a_2 \geq 0$ to the first and second equations, respectively, and a dummy objective function which minimizes the sum of these artificial variables in the problem, we get:

$$Min\ z = a_1 + a_2$$

Subject to:

$$10x_1 + 4x_2 + a_1 = 20$$
$$4x_1 + 3x_2 + a_2 = 10$$
$$x_1, x_2, a_1, a_2 \geq 0 \qquad\qquad\qquad \dots(4.19)$$

Taking a_1 and a_2 as the initial basic variables, (4.19) can be rewritten as:

$$10x_1 + 4x_2 + a_1 = 20$$
$$4x_1 + 3x_2 + a_2 = 10$$
$$z + 14x_1 + 7x_2 = 30$$
$$x_1, x_2, a_1, a_2 \geq 0$$

Basic Variables	x_1	x_2	a_1	a_2	Solution	Ratio
a_1	10	4	1	0	20	2
a_2	4	3	0	1	10	2.5
z	14	7	0	0	30	

Basic Variables	x_1	x_2	a_1	a_2	Solution	Ratio
x_1	1	0.4	0.1	0	2	5
a_2	0	1.4	−0.4	1	2	1.428571
z	0	1.4	−1.4	0	2	

Basic Variables	x_1	x_2	a_1	a_2	Solution
x_1	1	0	0.214	− 0.286	1.429
x_2	0	1	− 0.286	0.714	1.429
z	0	0	−1	−1	0

Now that all coefficients of zero are $z \leq 0$, the optimality condition is satisfied. Since a_1 and a_2 are the starting basic variables, therefore the inverse of $\mathbf{A}$ is $\begin{bmatrix} 0.214 & -0.286 \\ -0.286 & 0.714 \end{bmatrix}$.

Example 4.12. *Use the simplex method to find the inverse of* $A = \begin{bmatrix} 3 & 2 \\ 1 & 1 \end{bmatrix}$.

Solution: Let us consider the following system of equations:

$$\begin{bmatrix} 3 & 2 \\ 1 & 1 \end{bmatrix}\begin{bmatrix} x_1 \\ x_2 \end{bmatrix} = \begin{bmatrix} 9 \\ 6 \end{bmatrix} \qquad \qquad ...(4.20)$$

where $b = \begin{bmatrix} 9 \\ 6 \end{bmatrix}$ is a dummy column vector. Introducing the artificial variables a_1 and a_2 to the first and second equations, respectively, and a dummy objective function which minimizes the sum of these artificial variables in the problem, we get

$$\text{Min } z = a_1 + a_2$$

Subject to

$$3x_1 + 2x_2 + a_1 = 9$$
$$x_1 + x_2 + a_2 = 6 \qquad \qquad ...(4.21)$$
$$x_1, x_2, a_1, a_2 \geq 0$$

Taking a_1 and a_2 as the initial basic variables, (4.21) can be rewritten as:

$$3x_1 + 2x_2 + a_1 = 9$$
$$x_1 + x_2 + a_2 = 6$$
$$z + 4x_1 + 3x_2 = 15$$
$$x_1, x_2, a_1, a_2 \geq 0$$

Basic Variables	x_1	x_2	a_1	a_2	Solution	Ratio
a_1	3	2	1	0	9	3
a_2	1	1	0	1	6	6
z	4	3	0	0	15	

Basic Variables	x_1	x_2	a_1	a_2	Solution	Ratio
x_1	1	0.666667	0.333333	0	3	4.5
a_2	0	0.333333	− 0.33333	1	3	9
z	0	0.333333	− 1.33333	0	3	

Basic Variables	x_1	x_2	a_1	a_2	Solution	Ratio
x_2	1.5	1	0.5	0	4.5	
a_2	− 0.5	0	− 0.5	1	1.5	
z	− 0.5	0	− 1.5	0	1.5	

Since all the values of $z \leq 0$, the optimum solution has been reached. But **A** is still not converted to a unit matrix. In order to do this, we allow x_1 to enter the basis and drop a_2 from the basis.

Basic Variables	x_1	x_2	a_1	a_2	Solution	Ratio
x_2	0	1	−1	3	9	
x_1	1	0	1	−2	−3	
z	0	0	−1	−1	0	

Again, we have $z \leq 0$, that is, the optimality condition is still not violated. But it should be noted that the feasibility in this case is violated. Since a_1 and a_2 are the starting variables, therefore the inverse of **A** is $\begin{bmatrix} 1 & -2 \\ -1 & 2 \end{bmatrix}$ (rows have been rearranged).

4.8. Special Cases

4.8.1. The Problem with Alternative or Multiple Solutions

This case with the graphical method has already been discussed in Section 2.8.1. In the simplex calculation, if in the final table, the value of z for any non-basic variable takes a zero value, then it is an indication that the problem has an alternative solution. In this situation, we allow this non-basic variable with a value zero in the z row to enter the basis, and after deciding the leaving variable from the basis by the usual criterion, we determine a new optimal basic feasible solution with the same value of objective function as with the earlier optimal basic feasible solution.

Example 4.13. *Use the simplex method to solve Example 2.9.*

Solution: Writing the problem into its standard form, we have:

$$\text{Maximize } z = x_1 + 2x_2$$

Subject to the constraints:

$$6x_1 + 4x_2 + s_1 = 24$$
$$x_1 + 2x_2 + s_2 = 6$$
$$-x_1 + x_2 + s_3 = 1$$
$$x_2 + s_4 = 2$$
$$x_1, x_2, s_1, s_2, s_3, s_4 \geq 0$$

Rewriting the previous problem, we have:

$$6x_1 + 4x_2 + s_1 = 24$$
$$x_1 + 2x_2 + s_2 = 6$$
$$-x_1 + x_2 + s_3 = 1$$
$$x_2 + s_4 = 2$$
$$z - x_1 - 2x_2 = 0, \, x_1, x_2, s_1, s_2, s_3, s_4 \geq 0$$

Basic Variables	x_1	x_2	s_1	s_2	s_3	s_4	Solution	Ratio
s_1	6	4	1	0	0	0	24	6
s_2	1	2	0	1	0	0	6	3
s_3	−1	1	0	0	1	0	1	1
s_4	0	1	0	0	0	1	2	2
z	−1	−2	0	0	0	0	0	

Basic Variables	x_1	x_2	s_1	s_2	s_3	s_4	Solution	Ratio
s_1	10	0	1	0	−4	0	20	2
s_2	3	0	0	1	−2	0	4	1.33
x_2	−1	1	0	0	1	0	1	
s_4	1	0	0	0	−1	1	1	1
z	−3	0	0	0	2	0	2	

Basic Variables	x_1	x_2	s_1	s_2	s_3	s_4	Solution	Ratio
s_1	0	0	1	0	6	−10	10	1.666667
s_2	0	0	0	1	1	−3	1	1
x_2	0	1	0	0	0	1	2	
x_1	1	0	0	0	−1	1	1	
z	0	0	0	0	−1	3	5	

Basic Variables	x_1	x_2	s_1	s_2	s_3	s_4	Solution	Ratio
s_1	0	0	1	− 6	0	8	4	0.5
s_3	0	0	0	1	1	−3	1	
x_2	0	1	0	0	1	1	2	2
x_1	1	0	0	1	0	−2	2	
z	0	0	0	1	0	0	6	

Since all the elements in the z row are ≥ 0, the optimality condition is now satisfied. Hence, the optimum basic feasible solution to the given problem is $x_1 = 2$, $x_2 = 2$, and the corresponding maximum value of z is 6.

But the value of z for the non-basic variable s_4 is zero; this is an indication of the existence of an alternative optimum solution to the given problem. To find out the alternative optimum basic feasible solution to this problem, we allow s_4 to enter the basis and, accordingly, s_1 leaves the basis. The resulting table is shown as follows:

Basic Variables	x_1	x_2	s_1	s_2	s_3	s_4	Solution
s_4	0	0	0.125	− 0.75	0	1	0.5
s_3	0	0	3	− 17	1	21	13
x_2	0	1	− 0.125	0.75	0	0	1.5
x_1	1	0	0.25	− 0.5	0	0	3
z	0	0	0	1	0	0	6

Again, the optimality condition is still satisfied, and we have totally different values of x_1 and x_2, giving the same value of z.

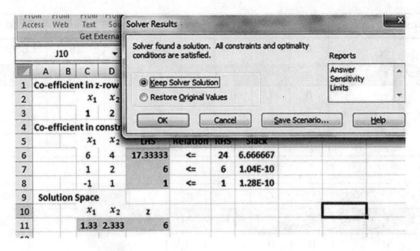

Fig. 4.7

Remark: In real life, alternative optima are very useful; they allow the user to select different solutions without experiencing any deterioration in the value of the objective function.

4.8.2. Unbounded Solutions

The case of unbounded solutions using the graphical method was discussed in Section 2.8.2. In the simplex calculation, the problem will be said to have an unbounded solution if all the numbers in the pivot column are ≤ 0.

Example 4.14. *Use the simplex method to solve* **Example 2.10**.

Solution: Write the given problem into its standard form.

$$\text{Maximize } z = 4x_1 + 10x_2$$

Subject to the constraints:

$$x_1 + s_1 = 10$$
$$x_1 + 2x_2 - s_2 = 6$$
$$x_1, x_2, s_1, s_2 \geq 0$$

By introducing an artificial variable $a \geq 0$ to the second constraint to have the initial basic feasible solution, the canonical form of the given problem can be written as

$$x_1 + s_1 = 10$$
$$x_1 + 2x_2 - s_2 + a = 6$$
$$z - (M + 4)x_1 - (2M + 10)x_2 + Ms_2 = -6M$$
$$x_1, x_2, s_1, s_2 \geq 0$$

Now the canonical form to the given problem is:

Basic Variables	x_1	x_2	s_1	s_2	a	Solution	Ratio
s_1	1	0	1	0	0	10	
a	1	2	0	-1	1	6	3
z	$-(M+4)$	$-(2M+10)$	0	M	0	$-6M$	

Basic Variables	x_1	x_2	s_1	s_2	Solution	Ratio
s_1	1	0	1	0	10	
x_2	0.5	1	0	-0.5	3	
z	1	0	0	-5	30	

This tableau is not optimal. However, at this point we are unable to perform further iterations, because as we attempt to carry out a ratio test with s_2 as the entering variable, it turns out that there is no ratio to compute. What this means is that as we attempt to bring s_2 in as a basic variable, none of the constraints will stop us from increasing its value to infinity. Now, as the value of s_2 increases, the objective function value will also increase correspondingly at a rate of 5. It follows that the problem does not have a finite optimal solution.

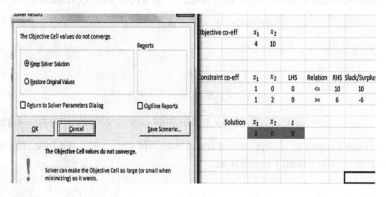

Fig. 4.8

It can also be seen from Figure 4.8 that the objective function values of the given problem do not converge, indicating that the problem has an unbounded solution.

4.8.3. The Problem with Inconsistent Constraint Equations

This case with the graphical method was discussed in Section 2.8.3. Here we will study, how to identify the condition in the simplex table under which simplex algorithm gives an indication of no solution to the given problem due to either inconsistency of the constraints or absence of any feasible solution. Inconsistency or infeasibility will never occur if all the constraints are of a $\leq$ type. For the other type of constraints where, the artificial variables are added to get the initial basic feasible solution, when artificial variables occur at positive level in the optimal basic feasible solution (and cannot be removed from the basis by replacing it with other legitimate variables, in any way) and are an indication that the given LP problem has no solution, this may be because of an the occurrence of an infeasible solution.

Example 4.15. *Use the simplex method to solve Example 2.11.*

Solution: Writing the problem in its standard form, we have:

$$Maximize\ z = x_1 + 4x_2$$

Subject to the constraints:

$$x_1 + 2x_2 - s_1 = 40$$
$$x_1 + x_2 + s_2 = 15$$
$$x_1, x_2, s_1, s_2 \geq 0$$

Introducing the artificial variable $a \geq 0$, in the first constraint we have

$$Maximize\ z = x_1 + 4x_2 - Ma$$

Subject to the constraints:

$$x_1 + 2x_2 - s_1 + a = 40$$
$$x_1 + x_2 + s_2 = 15$$

$x_1, x_2, s_1, s_2, a \geq 0$ and M is a large positive real number.

Taking a and s_2 as the initial basic feasible variables, we have the canonical form to the given problem as:

$$x_1 + 2x_2 - s_1 + a = 40$$

$$x_1 + x_2 + s_2 = 15$$

$$z - (1 + M)x_1 - (4 + 2M)x_2 + Ms_1 = -40M$$

$x_1, x_2, s_1, s_2 a \geq 0$ and M is a large positive real number.

Basic Variables	x_1	x_2	s_1	s_2	a	Solution	Ratio
a	1	2	−1	0	1	40	20
s_2	1	1	0	1	0	15	15
z	− (1 + M)	− (4 + 2M)	M	0	0	− 40M	

Basic Variables	x_1	x_2	s_1	s_2	a	Solution	Ratio
a	−1	0	−1	−2	1	10	
x_2	1	1	0	1	0	15	
z	3 + M	0	M	4 + 2M	0	60−10M	

Since all the elements in the z row are ≥ 0, the optimality condition is now satisfied. But the artificial variable "a" is still there in the basis at a positive level, indicating that the given LP is infeasible.

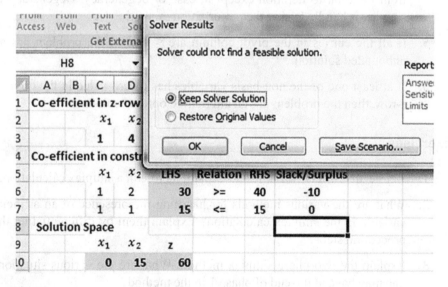

Fig. 4.9

From Figure 4.9 it can be seen that the Solver could not find a feasible solution to the given problem indicating that the given problem is infeasible.

Remarks: The following points are to be noted while making simplex calculations.

1. An artificial variable is introduced in the simplex calculation only to provide us an identity basis matrix to start the simplex calculation.

2. The presence of an artificial variable in the problem is handled either by the **M-Charnes** method or the two-phase method.

3. The presence of artificial variables in the final simplex table (when the optimality condition is satisfied) will give us one of the following two cases:

    (*i*) At least one artificial variable appears in the optimum basis at a positive level. In this case, the given problem does not have any feasible solution, possibly because the constraint equations are inconsistent.

    (*ii*) At least one artificial vector appears in the optimum basis at a zero level. Go to phase 2 but there may be redundancy in the original constraint equations.

4. The value of the objective function should always show the improvement from iteration to iteration except in case of degeneracy. Degeneracy in simplex is handled using the perturbation method.

5. If all the entries in the pivot column are ≤ 0, then the problem has an unbounded solution.

6. If at least one of the non-basis variables has a zero value in the optimal z-row, then the problem has an alternative optimal solution.

Exercises

1. Define an artificial variable and explain its role in a simplex calculation.

2. What are the various methods for handling the presence of an artificial variable in the simplex calculation? Explain them by differentiating the procedural steps.

3. Explain the two-phase simplex method. What are the various situations one may have at the end of phase 1 in the method?

4. Explain the steps involved in a Big M method for dealing with the presence of an artificial variable in a simplex calculation.

5. Explain the concepts of "degeneracy" and "cycling" in linear programming problems.

6. What is degeneracy in LPPs? What are the problems caused by degeneracy? How can this be resolved?

7. Explain the perturbation technique to handle degeneracy in LPPs.

8. Explain the existence of the following cases in the a simplex calculation:

   (*i*) Problem with alternative or multiple solutions

   (*ii*) Unbounded solution

   (*iii*) Problem with inconsistent constraint equations

9. Use the penalty simplex method to solve the following LPP:

   (*i*) $\qquad$ $Min\ z = x_1 + 2.5x_2 + x_3$

   Subject to

   $$1.5x_1 + 4x_2 + x_3 \leq 50$$
   $$x_1 + 2x_2 - x_3 \geq 15$$
   $$x_1 + 1.5x_2 + 2x_3 \leq 35$$
   $$x_1, x_2, x_3 \geq 0$$

   (*ii*) $\qquad$ $Max\ z = 10x_1 + 15x_2 + 23.5x_3$

   Subject to

   $$2x_1 - x_2 + 5x_3 \leq 50$$
   $$4x_1 - 0.5x_2 + 6x_3 \geq 75$$
   $$x_1, x_2, x_3 \geq 0$$

   (*iii*) $\qquad$ $Min\ z = 2x_1 + 3x_2 + 2x_3$

   Subject to

   $$2x_1 - 2x_2 + 2x_3 \leq 100$$
   $$2x_1 - 2x_2 + x_3 \geq 80$$
   $$2x_1 - 2x_2 + 2x_3 \geq 75$$
   $$x_1, x_2, x_3 \geq 0$$

(iv) $\qquad Min\ z = 10x_1 + 15x_2 + 23.5x_3$

Subject to

$$2x_1 - x_2 + 5x_3 \leq 50$$
$$4 - 0.5x_2 + 6x_3 \geq 75$$
$$x_1, x_2, x_3 \geq 0$$

(v) $\qquad Max\ z = x_1 + x_2 + x_3$

Subject to

$$3x_1 - x_2 + 2x_3 = 50$$
$$x_1 + 2x_2 + x_3 = 30$$
$$2x_1 - 1.5x_2 + x_3 = 35$$
$$x_1, x_2, x_3 \geq 0$$

(vi) $\qquad Min\ z = 2x_1 + 3x_2 + 3x_3$

Subject to

$$x_1 + 2x_2 + 2x_3 \leq 100$$
$$-x_1 - 2x_2 + x_3 \leq -80$$
$$-3x_1 - 2x_2 + 2x_3 \geq 60$$
$$x_2, x_3 \geq 0 \text{ and } x_1 \text{ is unrestricted in sign.}$$

(vii) $\qquad Max\ z = 3x_1 + 2x_2$

Subject to

$$x_1 + 2x_2 \leq 20$$
$$4x_1 + x_2 \leq 30$$
$$3x_1 - x_2 \geq 0$$
$$x_1, x_2, \geq 0$$

10. Solve the following LPP:

$$Max\ z = 10x_1 + 30x_2 + 30x_3$$

Subject to

$$x_1 + 4x_2 + 2x_3 \leq 90$$
$$3x_1 - 3x_2 + x_3 \leq 100$$
$$x_1, x_2, x_3 \geq 0$$

From the optimal table of the previous problem, check for the existence of an alternative optima. If it exists, find another optimum solution to the same problem.

11. Use the two-phase method to solve the following LPP:

$$Max \ z = 2x_1 - 4x_2 + 2x_3$$

Subject to

$$2x_1 + 3x_2 + x_3 \geq 50$$
$$x_1 - 3x_2 + x_3 \geq 30$$
$$x_1 + 2x_2 + 4x_3 \geq 45$$
$$x_1, x_2, x_3 \geq 0$$

Does the problem have a unique solution? If not, find an alternative optimal solution to the given problem.

12. Solve the following system of simultaneous linear equations using the simplex method:

(*i*) $x_1 + 2x_2 = 40; \ x_1 + x_2 = 25$

(*ii*) $4x_1 + 3x_2 = 10; \ 5x_1 + 6x_2 = 15$

(*iii*) $x_1 + x_2 + x_3 = 40; \ x_1 - 2x_2 + 1.5x_3 = 25; \ 2x_1 + x_2 + 0.5x_3 = 35;$ and $x_1,$
     $x_2, x_3, \geq 0$

13. Find the inverse of the following matrix using the simplex method:

(*i*) $\begin{bmatrix} 2 & 1 \\ 3 & 1 \end{bmatrix}$

(*ii*) $\begin{bmatrix} -1 & 1 \\ 1 & 2 \end{bmatrix}$

(*iii*) $\begin{bmatrix} 4 & 5 & 1 \\ 1 & 4 & 3 \\ 1 & 8 & 7 \end{bmatrix}$

CHAPTER 5

DUALITY

5.1. Introduction

The overall concept of duality is based on the idea that "every linear programming problem associates with another linear programming problem" which is called its "dual," so that both the problems are *dual* for one another. In this, if we can solve one problem, then the solution of the other problem (its dual) can be found from the solution of the original problem. This concept has its applications from a computational point of view. Sometimes when some linear programming problems involve a large number of constraints or have a large number of constraints of a ≥ or = type, then finding the dual of these types of problems and solving them to get the solution of the original problems will be easier and require less computational time and effort as compared to solving the original problem.

5.2. Rules for Finding the Dual of a Given Linear Programming Problem

The following are the rules for writing a dual to a given linear programming problem:

(i) If the objective function of the primal problem is maximization, then the dual of it will have minimization as its objective function and vice versa.

(ii) Inequalities of the primal problem get reversed in the dual.

(iii) Objective function coefficients of the primal become RHS coefficients of the dual constraint and vice versa.

(iv) The number of constraints in the primal is equal to the number of variables in the dual, and the number of variables in the primal is equal to the number of constraints in the dual.

(v) The column corresponding to each variable in the primal becomes a constraint in the dual, and the row coefficient in a constraint becomes the column coefficients of the dual variable (associated with that constraint) in the dual problem.

(vi) The dual variables corresponding to inequalities of type ≤ in primal maximization problems become non-negative variables in the dual problem, whereas the dual variables corresponding to inequalities of type ≤ in primal minimization problems become non-positive variables in the dual problem.

(vii) The dual variables corresponding to inequalities of type ≥ in primal maximization problems have non-positive value in the dual problem, whereas the dual variables corresponding to inequalities of type ≥ in primal minimization problems have non-negative values in the dual problem.

(viii) The dual variables corresponding to = (equality) constraints in the primal problem become unrestricted in sign in the dual problem.

(ix) Any dual constraint corresponding to a primal variable which is unrestricted in sign will have equality in the dual problem.

(x) A dual constraint corresponding to a non-negative primal variable will have ≤ inequality in a maximization problem and ≥ inequality in a minimization problem.

(xi) A dual constraint corresponding to a non-positive primal variable will have ≥ inequality in a maximization problem and ≤ inequality in a minimization problem.

The previous rules are summarized in the following table:

Primal	*Dual*
Maximize	Minimize
Number of constraints	Number of variables
Number of variables	Number of constraints
j^{th} constraint ≤	j^{th} variable ≥ 0

j^{th} constraint $\geq$	j^{th} variable ≤ 0
j^{th} constraint $=$	j^{th} variable unrestricted
i^{th} variable ≥ 0	i^{th} constraint $\geq$
i^{th} variable ≤ 0	i^{th} constraint $\leq$
i^{th} variable unrestricted	i^{th} constraint $=$

Consider the general linear programming problem with m constraints in n variables:

Maximize $\quad Z = c_1 x_1 + c_2 x_2 + \cdots \ldots \ldots \ldots \ldots \ldots + c_n x_n$

Subject to the constraints

$$a_{11} x_1 + a_{12} x_2 + \cdots \ldots \ldots \ldots \ldots + a_{1n} x_n \leq b_1$$
$$a_{21} x_1 + a_{22} x_2 + \cdots \ldots \ldots \ldots \ldots + a_{2n} x_n \leq b_2 \quad \text{(Primal)}$$
$$\vdots$$
$$\vdots$$
$$a_{m1} x_1 + a_{m2} x_2 + \cdots \ldots \ldots \ldots \ldots + a_{mn} x_n \leq b_m$$
$$x_1, x_2, \ldots \ldots x_n \geq 0 \qquad \qquad \text{...(5.1)}$$

The primal can also be written in the form

Maximize $\quad Z = \sum_{j=1}^{n} c_j x_j$

Subject to the constraints

$$\sum_{j=1}^{n} a_{ij} x_j \leq b_i, x_j \geq 0, \quad j = 1, 2, \ldots n, \quad i = 1, 2, \ldots m \qquad \text{...(5.2)}$$

Let $w_1, w_2, \ldots \ldots w_m$ be the m dual variables associated with the m constraints of the primal problem such that the dual of the given primal problem can be written as:

Minimize $\quad W = b_1 w_1 + b_2 w_2 + \cdots \ldots \ldots \ldots + b_m w_m$

Subject to the constraints

$$a_{11} w_1 + a_{21} w_2 + \cdots \ldots \ldots \ldots \ldots + a_{m1} w_m \leq c_1$$
$$a_{12} w_1 + a_{22} w_2 + \cdots \ldots \ldots \ldots \ldots + a_{m2} w_m \leq c_2 \quad \text{(Dual)}$$
$$\vdots$$
$$\vdots$$
$$a_{1n} w_1 + a_{2n} w_2 + \cdots \ldots \ldots \ldots + a_{mn} w_m \leq c_n$$
$$w_1, w_2, \ldots \ldots w_m \geq 0 \qquad \qquad \text{...(5.3)}$$

or   Minimize $w = \sum_{i=1}^{m} b_i w_i$

Subject to the constraints

$\sum_{i=1}^{m} a_{ij} w_i \geq c_j, \quad j = 1, 2, \dots n, \quad w_i \geq 0, \quad i = 1, 2 \dots m$ ...(5.4)

Note: Any given linear programming problem is said to be in "symmetric form" if all variables are non-negative and all constraints have inequalities of a ≤ type in the case of maximization and in minimization problems, they are of a ≥ type. The previous primal and dual problems are in symmetric form.

The primal and dual problems in **asymmetric form** can be written as

(Primal)

Maximize $Z = \sum_{j=1}^{n} c_j x_j$

Subject to the constraints

$\sum_{j=1}^{n} a_{ij} x_j \leq b_i, i = 1, 2 \dots \dots k < m$

$\sum_{j=1}^{n} a_{ij} x_j = b_i, i = k+1, k+2, \dots \dots m$

$x_j \geq 0, j = 1, 2 \dots \dots p < n,$

x_j unrestricted in sign, $j = p+1,$

$p+2, \dots \dots n$ ...(5.5)

(Dual)

Minimize $W = \sum_{i=1}^{m} b_i w_i$

Subject to the constraints

$\sum_{i=1}^{m} a_{ij} w_i \geq c_j, \quad j = 1, 2 \dots p < n$

$\sum_{i=1}^{m} a_{ij} w_i = c_j, j = p+1, p+2, \dots \dots n$

$w_i \geq 0, i = 1, 2 \dots \dots k,$

is w_i unrestricted in sign, $i = k+1, k+2 \dots \dots m$ ...(5.6)

Example 5.1. (Symmetric form): *Write the dual of the following LPP:*

$$Max \ Z = x_1 + 2x_2 + 3x_3$$

Subject to

$$3x_1 + x_2 \leq 70$$

$$-x_1 + x_2 + 4x_3 \leq 30$$

$$2x_1 - x_2 + 3x_3 \leq 60$$

$$x_1, x_2, x_3 \geq 0$$

Solution: Let w_1, w_2, and w_3 be the dual variables (corresponding to three primal constraints). Then the dual of the given problem is:

$$\text{Min } W = 70w_1 + 30w_2 + 60w_3$$

Subject to

$$3w_1 - w_2 + 2w_3 \geq 1$$

$$w_1 + w_2 - w_3 \geq 2$$

$$4w_2 + 3w_3 \geq 3$$

$$w_1, w_2, w_3 \geq 0$$

The previous can also be done using matrix notation:

Primal Problem: $\text{Max } Z = (1\,2\,3) \begin{pmatrix} x_1 \\ x_2 \\ x_3 \end{pmatrix}$

Subject to

$$\begin{pmatrix} 3 & 1 & 0 \\ -1 & 1 & 4 \\ 2 & -1 & 3 \end{pmatrix} \begin{pmatrix} x_1 \\ x_2 \\ x_3 \end{pmatrix} \leq \begin{pmatrix} 70 \\ 30 \\ 60 \end{pmatrix}$$

$$x_1, x_2, x_3 \geq 0$$

Dual Problem: $\text{Min } W = (70\,30\,60) \begin{pmatrix} w_1 \\ w_2 \\ w_3 \end{pmatrix}$

Subject to

$$\begin{pmatrix} 3 & -1 & 2 \\ 1 & 1 & -1 \\ 0 & 4 & 3 \end{pmatrix} \begin{pmatrix} w_1 \\ w_2 \\ w_3 \end{pmatrix} \geq \begin{pmatrix} 1 \\ 2 \\ 3 \end{pmatrix}$$

$$w_1, w_2, w_3 \geq 0$$

Example 5.2. *(Asymmetric problem):* Write the dual of the following LPP:

Maximize $\quad Z = 13x_1 + 7x_2$

Subject to the constraints

$$5x_1 + 8x_2 \leq 15$$
$$2x_1 + 5x_2 \geq 20$$
$$x_1 + 9x_2 = 25, x_1 \text{ unrestricted in sign, } x_2 \geq 0$$

Solution: Let w_1, w_2, and w_3 be the dual variables associated with three primal constraints, and then the dual of the given problem is:

$$\text{Minimize } 15w_1 + 20w_2 + 25w_3$$

Subject to the constraints

$$5w_1 + 2w_2 + w_3 = 13$$
$$8w_1 + 5w_2 + 9w_3 \geq 7$$
$$w_1 \geq 0, w_2 \leq 0, w_3 \text{ unrestricted in sign.}$$

(Given previously is an example of an asymmetric problem. The problem has three constraints in two variables, which means its dual will have three variables and two constraints. Also, it should be noted that the second primal constraint has a $\geq$ inequality, so the corresponding dual variable (w_2) should be ≤ 0. Since the first primal variable is unrestricted in sign, the corresponding dual constraint (first dual constraint) should have an equality sign. Finally, the last primal constraint has an equality sign, and hence the corresponding dual variable should be unrestricted in sign).

Example 5.3. *Write the dual of the following LPP:*

$$\text{Minimize } W = 7w_1 + 5w_2$$

Subject to the constraints

$$w_1 + 3w_2 \geq 5$$
$$0.5w_1 + w_2 \geq 3$$
$$w_1 - 0.5w_2 \geq 6, \ w_1, w_2 \text{ unrestricted in sign}$$

Solution: Let x_1, x_2, and x_3 be the dual variables, and then the dual of the given problem is:

Maximize $\qquad Z = 5x_1 + 3x_2 + 6x_3$

Subject to the constraints

$$x_1 + 0.5x_2 + x_3 = 7$$
$$3x_1 + x_2 - 0.5x_3 = 5, x_1, x_2, x_3 \geq 0$$

Example 5.4. *Write the dual of the following LPP:*

$$\text{Maximize } Z = 6x_1 + 7x_2$$

Subject to the constraints

$$2x_1 + 3x_2 = 7$$
$$-2x_1 + 5x_2 \geq 3$$
$$4x_1 + 7x_2 \leq 10, x_1 \geq 0, x_2 \text{ unrestricted in sign}$$

Solution: Let w_1, w_2, and w_3 be the dual variables and then the dual of the given problem is:

$$\text{Minimize } W = 7w_1 + 3w_2 + 10w_3$$

Subject to the constraints

$$2w_1 - 2w_2 + 4w_3 \geq 6$$
$$3w_1 + 5w_2 + 7w_3 = 7$$

w_1 unrestricted in sign, $w_2 \leq 0$, $w_3 \geq 0$

5.3. Finding the Optimal Dual Solution from the Optimal Table of the Primal Problem

The following table gives the layout of the simplex table discussed in Chapters 3 and 4.

Basic Variables	Initial non-basic variables	Initial basic variables	Solution column
Space for Basic variables	Columns for constraint coefficients	**Identity Matrix**	
Z-row			

(Initial Table)

Basic Variables	Initial non-basic variables	Initial basic variables	Solution column
Space for Basic variables	Columns for constraint coefficients	**Basis Inverse Matrix**	
Z-row			

(Any other table including the optimal)

As mentioned in an earlier chapter, simplex calculations start with a set of basic variables whose constraint coefficients always give us an identity matrix. The information of the basis inverse at any stage of the calculation is very important, as it is the key to computing all other elements of the associated table. Given the optimal table of the primal (dual) problem, the optimal solution of the other problem can be easily obtained by using any of the two following methods.

Method I. The optimal value of the ith dual variable will be obtained by taking the sum of the optimal z-coefficient of the ith starting primal basic variable and its original objective coefficient.

Method II. Optimal values of dual variables will be equal to the product of the row vector of the original objective coefficients of the optimal primal basic variables and the optimal primal basis inverse.

Example 5.5. *Use the primal optimal table to obtain the optimal dual solution of the following LP problem.*

$$\text{Maximize } z = 9x_1 - 6x_2 + 9x_3$$

Subject to the constraints

$$3x_1 - x_2 + 2x_3 \leq 28$$
$$x_1 - 2x_2 + 3x_3 \leq 30$$
$$5x_1 + x_2 + 2x_3 \leq 37$$
$$x_1 + x_2 + x_3 \leq 85$$
$$2x_1 + x_2 + x_3 \leq 18, \qquad x_1, x_2, x_3 \geq 0$$

Solution: The dual of the given problem is:

(Dual) Minimize $\qquad w = 28w_1 + 30w_2 + 37w_3 + 85w_4 + 18w_5$

Subject to the constraints

$$3w_1 + w_2 + 5w_3 + 1w_4 + 2w_5 \geq 9$$
$$-1w_1 - 2w_2 + w_3 + w_4 - w_5 \geq -6$$
$$2w_1 + 3w_2 + 2w_3 + w_4 + w_5 \geq 9$$
$$W_i \geq 0 \; \forall \; i$$

Rewriting the given primal problem in the canonical form, we have:

$$3x_1 - x_2 + 2x_3 + s_1 = 28$$
$$x_1 - 2x_2 + 3x_3 + s_2 = 30$$
$$5x_1 + x_2 + 2x_3 + s_3 = 37$$
$$x_1 + x_2 + x_3 + s_4 = 85$$
$$2x_1 - x_2 + x_3 + s_5 = 18$$
$$z - 9x_1 + 6x_2 - 9x_3 = 0$$
$$x_1, x_2, x_3, s_1, s_2, s_3, s_4, s_5 \geq 0$$

Basic	x_1	x_2	x_3	s_1	s_2	s_3	s_4	s_5	Solution	Ratio
s_1	3	−1	2	1	0	0	0	0	28	14
s_2	1	−2	3	0	1	0	0	0	30	10
s_3	5	1	2	0	0	1	0	0	37	18.5
s_4	1	1	1	0	0	0	1	0	85	85
s_5	2	−1	1	0	0	0	0	1	18	18
z	−9	6	−9	0	0	0	0	0	0	

(Initial table s_1, s_2, s_3, s_4, and s_5 as the starting basic variables)

Basic	x_1	x_2	x_3	s_1	s_2	s_3	s_4	s_5	Solution	Ratio
s_1	2.3333	0.3333	0	1	− 0.6666	0	0	0	8	3.428
x_3	0.3333	−0.6666	1	0	0.3333	0	0	0	10	30
s_3	4.3333	2.3333	0	0	− 0.6666	1	0	0	17	3.923
s_4	0.6666	1.6666	0	0	− 0.3333	0	1	0	75	112.5
s_5	1.6666	−0.3333	0	0	− 0.3333	0	0	1	8	4.8
z	− 6	0	0	0	3	0	0	0	90	

Basic	x_1	x_2	x_3	s_1	s_2	s_3	s_4	s_5	Solution
x_1	1	0.14286	0	0.4285	− 0.2857	0	0	0	3.4285
x_3	0	− 0.7142	1	− 0.1428	0.4285	0	0	0	8.8571
s_3	0	1.7142	0	− 1.8571	0.5714	1	0	0	2.1428
s_4	0	1.5714	0	− 0.2857	− 0.1428	0	1	0	72.714
s_5	0	− 0.5714	0	− 0.7142	0.1428	0	0	1	2.2857
z	0	0.8571	0	2.5714	1.2857	0	0	0	110.57

Since all the elements in the z row are ≥ 0, the optimality condition is now satisfied. Hence, the optimum solution for the primal problem is $x_1 = 3.42857$, $x_2 = 0$, $x_3 = 8.857$, $z = 110.571$.

Finding the optimal dual solution

Method I

Dual variables (i)	Primal starting basic variables (ii)	Optimal primal z-coefficient of starting basic variables (iii)	Original objective function coefficient of starting basic variables (iv)	Optimal value of dual variables (v = iii + iv)
w_1	s_1	2.5714	0	2.5714
w_2	s_2	1.2857	0	1.2857
w_3	s_3	0	0	0
w_4	s_4	0	0	0
w_5	s_5	0	0	0

Hence, the optimal dual solution is $w_1 = 2.57143$, $w_2 = 1.2857$, $w_3 = w_4 = w_5 = 0$, and the corresponding value of the objective function is 110.57.

Method II

Optimal values of dual variables will be equal to the product of the row vector of the original objective coefficients of the optimal primal basic variables and the optimal primal basis inverse.

$$\text{Optimal basis inverse, } \boldsymbol{B}^{-1} = \begin{pmatrix} 0.4285 & -0.2857 & 0 & 0 & 0 \\ -0.1428 & 0.4285 & 0 & 0 & 0 \\ -1.8571 & 0.5714 & 1 & 0 & 0 \\ -0.2857 & -0.1428 & 0 & 1 & 0 \\ -0.7142 & 0.1428 & 0 & 0 & 1 \end{pmatrix},$$

the row vector of the original objective function coefficients of the optimal primal basic variables $(x_1, x_2, s_3, s_4, s_5)$ is (9 9 0 0 0).

Hence, the optimal values of the dual variables $(w_1, w_2, w_3, w_4, w_5)$ are given by:

$$(9\,9\,0\,0\,0)*\begin{pmatrix} 0.4285 & -0.2857 & 0 & 0 & 0 \\ -0.1428 & 0.4285 & 0 & 0 & 0 \\ -1.8571 & 0.5714 & 1 & 0 & 0 \\ -0.2857 & -0.1428 & 0 & 1 & 0 \\ -0.7142 & 0.1428 & 0 & 0 & 1 \end{pmatrix} = (2.57143\ 1.2857\ 0\ 0\ 0)$$

Example 5.6. *Write the dual of the following LP problem:*

$$Maximize\ z = -x_1 - 2x_2$$

Subject to the constraints:

$$3x_1 + x_2 = 11$$
$$4x_1 + 3x_2 \geq 20$$
$$x_1 + 2x_2 \leq 14$$
$$x_1, x_2 \geq 0$$

And use the primal optimal table to find the optimal dual solution.

Solution: The dual of the given problem is:

$$Minimize\ w = 11w_1 + 20w_2 + 14w_3$$

Subject to the constraints:

$$3w_1 + 4w_2 + w_3 \geq -1$$
$$w_1 + 3w_2 + 2w_3 \geq -2$$

w_1 is unrestricted, $\quad w_2 \leq 0$ and $w_3 \geq 0$

The primal problem can be rewritten as:

maximize $\quad z = -x_1 - 2x_2 - Ma_1 - Ma_2$

Subject to the constraints

$$3x_1 + x_2 + a_1 = 11$$
$$4x_1 + 3x_2 - s_1 + a_2 = 20$$
$$x_1 + 2x_2 + s_2 = 14$$

$x_1, x_2, s_1, s_2, a_1, a_2 \geq 0$ and M is a large positive number.

Now the canonical form for the given primal problem is:

$$3x_1 + x_2 + a_1 = 11$$
$$4x_1 + 3x_2 - s_1 + a_2 = 20$$
$$x_1 + 2x_2 + s_2 = 14$$
$$z + (1 - 7M) x_1 + (2 - 4M) x_2 + Ms_1 = -31M$$
$$x_1, x_2, s_1, s_2, a_1, a_2 \geq 0$$

And the following is the optimal table for the primal problem.

Basic	x_1	x_2	s_1	s_2	a_1	a_2	Solution
x_1	1	0	0.2	0	0.6	-0.2	2.6
x_2	0	1	-0.6	0	-0.8	0.6	3.2
s_1	0	0	1	1	1	-1	5
z	0	0	1	0	$M+1$	$M-1$	-9

Where a_1, a_2, and s_2 are the starting initial basic variables, and hence the optimal basis inverse is:

$$B^{-1} = \begin{pmatrix} 0.6 & -0.2 & 0 \\ -0.8 & 0.6 & 0 \\ 1 & -1 & 1 \end{pmatrix}$$

Finding the optimal dual solution

Method I

Dual variables (i)	Primal starting basic variables (ii)	Optimal primal z-coefficient of starting basic variables (iii)	Original objective function coefficient of starting basic variables (iv)	Optimal value of dual variables (v = iii + iv)
w_1	a_1	$M+1$	$-M$	1
w_2	a_2	$M-1$	$-M$	-1
w_3	s_2	0	0	0

Hence, the optimal dual solution is $w_1 = 1$, $w_2 = -1$, and $w_3 = 0$, and the corresponding value of the objective function is -9.

Method II

Optimal values of dual variables will be equal to the product of the row vector of the original objective function coefficients of the optimal primal basic variables and the optimal primal basis inverse.

Optimal basis inverse, $\mathbf{B}^{-1} = \begin{pmatrix} 0.6 & -0.2 & 0 \\ -0.8 & 0.6 & 0 \\ 1 & -1 & 1 \end{pmatrix}$, row vector of the original

objective function coefficients of the optimal primal basic variables, (x_1, x_2, s_2) is $(-1\ -2\ 0)$.

Hence, the optimal values of the dual variables (w_1, w_2, w_3) are given by

$$(-1-2\ 0) * \begin{pmatrix} 0.6 & -0.2 & 0 \\ -0.8 & 0.6 & 0 \\ 1 & -1 & 1 \end{pmatrix} = (1-1\ 0)$$

5.4. Use of the Graphical Method for Finding the Optimal Dual Solution

Consider the following LPP given in Example 2.8:

$$\text{Maximize } z = 4x_1 + x_2$$

Subject to the constraints:

$$2x_1 + x_2 \leq 101$$
$$x_1 + x_2 \leq 80$$
$$x_1 \leq 40$$
$$x_1, x_2 \geq 0$$

Then, the dual of this problem is,

$$\text{Minimize } w = 101\,w_1 + 80w_2 + 40w_3$$

Subject to the constraints:

$$2w_1 + w_2 + w_3 \geq 4$$
$$w_1 + w_2 \geq 1$$
$$w_1, w_2, w_3 \geq 0$$

The following diagram gives the graphical solution of the primal problem.

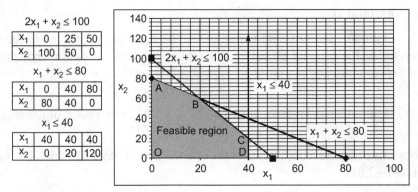

Fig. 5.1

$z^* = 180$ is the maximum value z occurring at point C (40, 20).

Solving the problem again by increasing the RHS of the first constraint by one unit, that is, by changing 100 to 101, we get:

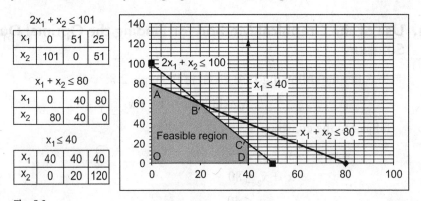

Fig. 5.2

Now, the new feasible region is denoted by $OAB'C'D$. Let z' be the new value of the objective function for the changed problem. Clearly, the maximum value of z' is 181, which occurred at C' (40, 21).

Here, the optimal value of the dual variable w_1 is the difference between the new maximum value of the primal objective function and the original maximum value of the primal objective function, that is, $w_1 = 1$. In other words, the dual variable represents the contribution in the objective function corresponding to one unit's worth of a resource. Similarly, to find the value of w_2 and w_3, the same procedure will be repeated with the second and third primal constraint, respectively.

Points to remember:

(i) The values of the dual variables at any iteration of the simplex method are called simplex multipliers.

(ii) The dual variables are also called shadow prices or imputed cost per unit of resources, representing worth per unit of resources and used to rank resources according to their contribution in the objective function.

(iii) Any LP problem can be considered as an input-output model, where resources are the input and profit obtained can be considered as output such that we try to maximize profit under limited resources, and the optimal solution cannot be reached unless profit is equal to the worth of the resources. As we move from one iteration to another, profit < worth of resources, and as soon as they become equal, we get the optimal solution.

Note. *Computation of optimal dual variables using* **Excel Solver** *will be discussed latter.*

5.5. Construction of a Simplex Table

A simplex table of any iteration can be easily generated using the original data, the basis inverse of the associated iteration, and the dual of the problem. Construction of the whole simplex table can be carried out by as follows:

I. Constraint coefficient columns including the solution column: At any stage of the simplex table calculations the constraint columns of the ith iteration can be obtained by taking the product of the basis inverse of the ith iteration and the original constraint coefficients.

II. Value of the z-row: At any stage of the simplex calculation, the z-row values of the variable $x_j \, \forall \, j$ can be calculated by taking the difference of the LHS and RHS of the jth dual constraint, that is, the z-row value for x_j = (LHS – RHS) of the jth constraint.

Example 5.7. *Consider the following LPP:*

$$Maximize \ z = x_1 + 4x_2 + x_3$$

Subject to

$$x_1 + x_2 - 2x_3 \leq 7$$
$$-3x_1 + x_2 + 2x_3 \leq 3$$
$$x_1, x_2, x_3 \geq 0$$

Consider the set of basic variables $= (x_4, x_2)$, *and the basis inverse* $= \begin{pmatrix} 1 & -1 \\ 0 & 1 \end{pmatrix}$.

Compute the entire simplex table associated with the given basic solution and

check for its optimality and feasibility. Here x_4 *is the slack variable associated with the first constraint.*

Solution: Rewriting the given problem into its standard form as follows:

Maximize $z = x_1 + 4x_2 + x3$

Subject to

$$x_1 + x_2 - 2x_3 + x_4 = 7$$
$$-3x_1 + x_2 + 2x_3 + x_5 = 3$$
$$x_1, x_2, x_3, x_4, x_5 \geq 0$$

(i) Column for x_1 = Basis inverse * (original constraint coefficient of x_1)

$$= \begin{pmatrix} 1 & -1 \\ 0 & 1 \end{pmatrix} * \begin{pmatrix} 1 \\ -3 \end{pmatrix} = \begin{pmatrix} 4 \\ -3 \end{pmatrix}$$

(ii) Column for x_2 = Basis inverse * (original constraint coefficient of x_2)

$$= \begin{pmatrix} 1 & -1 \\ 0 & 1 \end{pmatrix} * \begin{pmatrix} 1 \\ 1 \end{pmatrix} = \begin{pmatrix} 0 \\ 1 \end{pmatrix}$$

(iii) Column for x_3 = Basic inverse * (original constraint coefficient of x_3)

$$= \begin{pmatrix} 1 & -1 \\ 0 & 1 \end{pmatrix} * \begin{pmatrix} -2 \\ 2 \end{pmatrix} = \begin{pmatrix} -4 \\ 2 \end{pmatrix}$$

(iv) Column for x_4 = Basic inverse * (original constraint coefficient of x_4)

$$= \begin{pmatrix} 1 & -1 \\ 0 & 1 \end{pmatrix} * \begin{pmatrix} 1 \\ 0 \end{pmatrix} = \begin{pmatrix} 1 \\ 0 \end{pmatrix}$$

(v) Column for x_5 = Basic inverse * (original constraint coefficient of x_5)

$$= \begin{pmatrix} 1 & -1 \\ 0 & 1 \end{pmatrix} * \begin{pmatrix} 0 \\ 1 \end{pmatrix} = \begin{pmatrix} -1 \\ 1 \end{pmatrix}$$

(vi) Solution column, that is, the value of (x_4, x_2) = Basis inverse * (RHS of original constraints)

$$= \begin{pmatrix} 1 & -1 \\ 0 & 1 \end{pmatrix} * \begin{pmatrix} 7 \\ 3 \end{pmatrix} = \begin{pmatrix} 4 \\ 3 \end{pmatrix}$$

To find the z-row values, we need to write the dual of the given problem, as follows:

$$\text{Minimize } z = 7w_1 + 3w_2$$

Subject to

$$w_1 - 3w_2 \geq 1$$
$$w_1 + w_2 \geq 4$$
$$-2w_1 + 2w_2 \geq 1$$
$$w_1, w_2 \geq 0$$

The value of the dual variable at this stage, $(0 \quad 4) * \begin{pmatrix} 1 & -1 \\ 0 & 1 \end{pmatrix} = (0 \quad 4) = (w_1 \quad w_2)$

(i) Value of z-row for $\quad x_1 =$ (LHS – RHS) of the first dual constraints
$$= -3 * 4 - 1 = -12.$$

(ii) Value of z-row for $\quad x_2 =$ (LHS – RHS) of the second dual constraints
$$= 4 - 4 = 0.$$

(iii) Value of z-row for $\quad x_3 =$ (LHS – RHS) of the third dual constraints
$$= 2 * 4 - 1 = 7.$$

The following is the complete simplex table at this iteration:

Basic	x_1	x_2	x_3	x_4	x_5	Solution
x_4	4	0	– 4	1	– 1	4
x_2	– 3	1	2	0	1	3
z	–12	0	7	0	4	12

Since all the elements in the solution column are ≥ 0, the present basic solution is feasible but not optimal, since all elements in the z-row are not ≥ 0.

Note. *The following points are to be remembered while constructing a simplex table.*

(i) The z-row elements corresponding to the basic variables will always be equal to zero.

(ii) The columns corresponding to the basic variable will always provide an identity matrix.

5.6. Duality Theorems

Theorem 5.1: The dual of the dual is primal.

Proof: Let $\quad$ **Max** $f(x) = cx$

Subject to

$$Ax \leq b \qquad \qquad ...(5.7)$$

$x \geq 0$, be the given primal problem.

Where c is a row vector containing n elements, x is a column vector containing n decision variables, A is the coefficient matrix of dimension $m \times n$, and b is a column vector containing the RHS values of the constraints.

Then the dual corresponding to this primal problem is

$$\textbf{Min } f(y) = b'y$$

Subject to

$$A'y \geq c' \qquad \qquad ...(5.8)$$

$$y \geq 0,$$

Where b', A', and c' are the transposes of b, A, c, and y is an m-dimension column vector of the dual decision variables.

Rewriting (5.8), we have

$$\textbf{Max} - f(y) = - b'y$$

Subject to

$$-A'y \leq -c'$$

$$y \geq 0,$$

Now, the dual of the dual is

$$\text{Min} - f(w) = (- c')'w = - cw$$

Subject to

$$(-A')\,'w \geq (- b')' \Rightarrow - Aw \geq - b$$

$$w \geq 0,$$

Rewriting the previous, we have

$$\text{Max} f(w) = cw$$

Subject to

$$Aw \le b \qquad \qquad ...(5.9)$$

$w \ge 0$, where w is a column vector containing n decision variables of the dual of (5.8).

Now, we can clearly see that (5.7) and (5.9) are the same, and hence we can conclude that the dual of a dual is the primal.

Farkas Lemma: If A is any $m \times n$ real matrix and $b \in R^m$, then either

(*i*) $Ax = b, x \ge 0$, *or*

(*ii*) $w^T A \ge 0$ and $w^T b < 0$, $w \in R^m$ has a solution, but not both.

Proof: Here we need to show that we cannot have solutions to both (*i*) and (*ii*). On the contrary, let us assume that both have solutions and let x be a solution to (*i*) and w to (*ii*), so that we have $Ax = b, x \ge 0$ and $w^T A \ge 0$, and $w^T b < 0$.

Then, we have

$w^T A \ge 0 \Rightarrow w^T A x \ge 0$ (since $x \ge 0$) $\Rightarrow w^T b \ge 0$ (from (*i*)), which is a contradiction of (*ii*), and hence if (*i*) is true (*ii*) cannot be true.

Theorem 5.2 (Weak Duality Theorem): For any given feasible solution, the value of the objective function of the minimization problem (dual) is always greater than or equal to the value of the objective function of the maximization (primal) problem.

Proof: Let x_0 be any feasible solution to the following primal problem:

$$\text{Max} f(x) = cx$$

Subject to

$$Ax \le b \qquad \qquad ...(5.10)$$

$$x \ge 0$$

And w_0 be any feasible solution to the dual problem of (5.10); that is, w_0 is the feasible solution to

$$\textbf{Min} f(w) = b'w$$

Subject to

$$A'w \ge c' \qquad \qquad ...(5.11)$$

$$w \ge \mathbf{0},$$

Then, we have to show that $cx_0 \le b'w_0$

From the dual constraint, we have

$c' \le A'w_0$ (since w_0 is the feasible solution to the dual problem)

$\Rightarrow c \le w'_0 A \qquad \Rightarrow cx_0 \le w'_0 Ax_0 \qquad \Rightarrow cx_0 \le w'_0 b$

(Since x_0 is the feasible solution to the primal problem, we have $Ax_0 \le b$)

Hence, $\qquad cx_0 \le b'w_0 \qquad$ (since $b'w_0 = w'_0 b$)

Hence the theorem.

Theorem 5.3 (Optimality Theorem): If x_0 is the feasible solution to the following primal problem:

$$\text{Max } f(x) = cx$$

Subject to

$$Ax \le b$$

$$x \ge 0$$

And w_0 is a feasible solution to its dual, that is, w_0 is the feasible solution to

$$\text{Min } f(w) = b'w$$

Subject to

$$A'w \ge c'$$

$$w \ge 0,$$

And, if their objective function values are the same at these feasible solutions, then these are the optimal solutions of their respective problems.

Proof: *If x_0 and w_0 are the feasible solutions of* primal and dual problems, respectively, and if their objective functions are same at these solutions.

Then, we have

$$cx_0 = b'w_0$$

Further, let x^* and w^* be other feasible solutions of primal and dual problems, respectively, then by the weak duality theorem, we have

$$cx^* \le b'w_0 \qquad \Rightarrow cx^* \le cx_0 \qquad (\text{since } cx_0 = b'w_0) \; \forall x^*$$

Which proves that x_0 is an optimal solution of primal problem.

Similarly, we have

$$cx_0 \le b'w^* \Rightarrow b'w_0 \le b'w^*$$

And hence it shows that w_0 is an optimal solution for a dual.

Example 5.8. *Consider the following primal and dual linear programs.*

(Primal)   Maximize $Z = x_1 + 5x_2 + 3x_3$

Subject to the constraint

$$2x_1 + x_2 + 4x_3 \le 15$$
$$x_1 + 6x_2 + 5x_3 \le 20, \, x_1, x_2, x_3 \ge 0$$

(Dual)   Minimize $W = 15w_1 + 20\,w_2$

Subject to the constraint

$$2w_1 + w_2 \ge 1, \, w_1 + 6w_2 \ge 5, \, 4w_1 + 5w_2 \ge 3$$
$$w_1, w_2 \ge 0$$

Let $x_1 = 1$, $x_2 = 1$, $x_3 = 1$ be the feasible solution for the primal problem and $w_1 = 1$, $w_2 = 2$ be the feasible solution for the dual problem. Then the value of the objective function for the primal problem is $Z = 9$ and for the dual problem is $W = 55$. Since $(Z = 9) < (W = 55)$, it proves the weak duality theorem.

Theorem 5.4 (Complementary Slackness Theorem): Let x_0 be a feasible solution to the following primal problem:

$$\textbf{Max } f(x) = cx$$

Subject to

$$Ax \le b \qquad\qquad\qquad ...(5.12)$$
$$x \ge 0,$$

And w_0 is the feasible solution to the dual of (5.12), that is, w_0 is the feasible solution to

$$\textbf{Min } f(w) = b'w$$

Subject to

$$A'w \ge c' \qquad\qquad\qquad ...(5.13)$$
$$w \ge 0,$$

Then, the necessary and sufficient conditions for x_0 and w_0 to be optimal for their respective problems is $w'_0 (b - Ax_0) = 0$ and $x'_0 (A'w_0 - c') = \textbf{0}$.

Proof (Necessary condition): Let x_0 and w_0 be the optimal solutions of the primal and dual problems, respectively. Also let $p = w'_0 (b - Ax_0)$ and $q = x'_0 (A'w_0 - c')$.

Since $w'_0, (b - Ax_0), x'_0, (A'w_0 - c') \geq 0$, it implies $p, q \geq 0$.

Then, we have

$$p + q' = w'_0 b - w'_0 Ax_0 + w'_0 Ax_0 - cx_0 = w'_0 b - cx_0 = b'w_0 - cx_0 \ (b'w = w'b)$$

By Theorem (5.3), at the optimal points x_0 and w_0 the dual objective function is equal to the primal objective function, that is $b'w_0 = cx_0 \Rightarrow p + q' = 0$.

Hence, $p = 0$, $q = 0$; since $p \geq 0$ and $q \geq 0$, thus, the necessary condition is proved.

(Sufficient condition): Now, let us assume that $p = 0$, $q = 0$; then, we have to show x_0 and w_0 are the optimal solutions to the primal and dual problems, respectively.

Since $p + q' = 0$, we can write

$$b'w_0 - cx_0 = 0$$

$$\Rightarrow \qquad b'w_0 = cx_0$$

$\Rightarrow$ x_0 and w_0 are the optimal solutions to the primal and dual problems, respectively (by the optimality theorem).

Hence, the sufficient condition.

Note. For optimal feasible solutions of primal and dual problems, whenever inequality occurs in the ith constraint of either problem, then the corresponding i^{th} dual variable will have zero value. Similarly, if the ith variable is positive in either problem, then the corresponding constraint in its dual problem is having equality.

Example 5.9. *The following linear programming problem:*

$$Maximize \ z = 4x_1 + x_2$$

Subject to the constraints:

$$2x_1 + x_2 \leq 100$$

$$x_1 + x_2 \leq 80$$

$$x_1 \leq 40$$

$$x_1, x_2 \geq 0$$

has (40, 20) as an optimum solution. Use the complementary slackness condition to find the optimum solution of its dual.

Solution: Let w_1, w_2, and w_3 be the dual variables, and then the dual of the given problem is

$$\text{Minimize } w = 100w_1 + 80w_2 + 40w_3$$

Subject to

$$2w_1 + w_2 + w_3 \geq 4$$
$$w_1 + w_2 \geq 1$$
$$w_1, w_2, w_3 \geq 0$$

On evaluating the inequalities of the primal constraints at the optimal points $x_1^* = 40$ and $x_2^* = 20$, we find positive slack in the second constraint and zero slack in the first and third constraints. Thus, by the complementary slackness theorem, at the optimal point of the dual problem, we have $w_2^* = 0$.

Now, since x_1^* and x_2^* are non-zero, the dual constraints corresponding to these two variables will satisfy equality, and hence the dual surplus variables for these constraints should be equal to zero. Hence, we can write

$$2 w_2^* + w_3^* = 4 \text{ (first dual constraints with } w_2^* = 0, \text{ as explained previously)}$$

$$w_1^* = 1 \text{ (Second dual constraint)}$$

Clearly, $w_1^* = 1$, $w_2^* = 0$, and $w_3^* = 2$ is the optimal solution of the dual problem and the corresponding minimum value of the objective function is 180, which is same as the value of the primal objective function at the optimal solution $x_1^* = 40$ and $x_2^* = 20$.

Example 5.10. *Use the complementary slackness theorem to find the optimal solution of the following LPP:*

$$\text{Minimize } Z = 15x_1 + 9x_2 + 8x_3 + 2x_4$$

Subject to the constraints $10x_1 + 14x_2 + 6x_3 + x_4 \geq 300 \quad x_j \geq 0 \ \forall j$

Solution: The following is the dual of the given problem:

Maximize $W = 300 \ w$

Subject to the constraints

$10w \leq 15, \ 14w \leq 9, \ 6w \leq 8, \ w \leq 2, \qquad w \geq 0$

The solution of the dual problem is $w = \dfrac{9}{14}$, which is satisfied by all the constraints.

And the second constraint satisfies with equality, while other constraints satisfy strict inequality. The optimal value of objective function at this point is $W = 1350/7$.

Now, by the complementary slackness condition, the variable x_1 corresponding to the first dual constraint, which satisfies inequality, will be zero, and x_2 will be >0 because it is the dual variable corresponding to the second constraint which satisfies equality. Similarly, by following the same argument, we have:

$$x_1 = x_3 = x_4 = 0, \Rightarrow 14x_2 \geq 300 \Rightarrow x_2 \geq \frac{150}{7} \quad \text{(from the primal constraint)}$$

Now, since we have to minimize the value of $z = 9x_2$, it will be minimum at $x_2 = 150/7$, and hence Min $Z = 9 * \left(\dfrac{150}{7}\right) = \dfrac{1350}{7} = \text{Max } W$.

Example 5.11. *Suppose it has been claimed that (5, 35, 0) is the optimal solution of the primal problem given in Example 5.1. Verify the claim using the complementary slackness theorem.*

Solution: It can be seen that the given points satisfy the second primal constraint with equality, whereas the other two constraints are satisfied with strict inequality. Hence, by the complementary slackness theorem, w_1 and $w_3 = 0$. Now, since the first and second primal variables are different from zero, the first and second dual constraints should have equality, so that we have

$$3w_1 - w_2 + 2w_3 = 1$$
$$w_1 + w_2 - w_3 = 2 \quad \text{...(5.14)}$$

On putting with w_1 and $w_3 = 0$, we get $w_2 = -1$ and 2 from the two equations of (5.14), which are contradicting one another. Hence, (5, 35, 0) cannot be an optimal solution to the given primal problem.

Theorem 5.5 (Strong Duality Theorem): For the primal-dual relationship of a linear programming problem, if the primal (dual) problem has a finite optimal solution, then so does the dual (primal) problem, and their value of objective functions are same at the optimal point.

Proof: Let Maximize $z = \sum_{j=1}^{n} c_j x_j$

Subject to the constraints

$$\sum_{j=1}^{n} a_{ij} x_j \le b_i, \quad x_j \ge 0, \quad j = 1, 2 \dots n, \quad i = 1, 2 \dots m,$$

be the given primal problem and

Minimize $w = \sum_{i=1}^{m} b_i w_i$

Subject to the constraints

$$\sum_{i=1}^{m} a_{ij} w_i \ge c_j, w_i \ge 0, j = 1, 2 \dots n, i = 1, 2 \dots m$$

be the associated dual problem. Adding the slack variables $x_{n+} 1, x_n + 2,$ ..., $x_n +_m$ to the constraints of the primal problem, we have the following standard form:

Maximize $z = \sum_{j=1}^{n} c_j x_j^*$

Subject to the constraints

$$\sum_{j=1}^{n} a_{ij} x_j + x_{n+i} = b_i,$$

$$x_j \ge 0, j = 1, 2 \dots n,$$

and $x_{n+i} \ge 0, i = 1, 2 \dots m$ ...(5.15)

Assuming the non-negativity of the constants b_i, $\forall i$ and taking x_{n+1}, x_{n+2}, ..., x_{n+m} as the starting initial basic feasible variables, which constitute a basic feasible solution to the previous problem, (5. 15) is now in its canonical form, and hence the simplex calculation can be started.

Let $x*j, j = 1, 2, 3, \dots, n, (n + 1) \dots (n + m)$ be the resulting optimal solution of the primal problem.

Now, let $w*_i, i = 1, 2 \dots .. m$, be the values of dual variables associated with the optimal table of the primal problem. The primal problem has an optimal solution since its optimality condition is satisfied, and we know that the z-row coefficients of the optimal table of the primal problem can be written as the difference of the LHS and RHS of the dual constraints, and we can write:

$\overline{c}_j = \sum_{i=1}^{m} a_{ij} w_i * - c_j \ge 0, j = 1, 2, \dots, n$ (values of z-row coefficients corresponding to the given variables) ...(5.16)

$\overline{c}_{n+i} = w_i^* - 0 \ge 0, i = 1, 2, \dots, m$ (For the slack variables) ..(5.17)

(Refer to Sections 5.3 and 5.5)

Combining (5.16) and (5.17), we have

$$\sum_{i=1}^{m} a_{ij} w_i^* \ge c_j, j = 1, 2, \dots, n, \text{ and } w_i^* \ge 0, i = 1, 2, \dots, m$$

This implies that w_i^*, $i = 1, 2 \ldots, m$ constitutes a feasible solution to the dual problem.

Let us assume that when the simplex algorithm terminates, the equation of z reads

as, $z = z^* - \sum_{k=1}^{n+m} \bar{c}_k x_k$ ...(5.18)

At the optimal point the value of $\bar{c}_k$ is non-negative, and in fact $\bar{c}_k$ has a zero value for the basic variables and positive values for the remaining and the maximum value of the objective function, and z^* at this point is given by

$z = \sum_{j=1}^{n} c_j x_j^*$ ...(5.19)

Rewriting (5.18), we have

$$z = z^* - \sum_{k=1}^{n} \bar{c}_k x_k - \sum_{i=1}^{m} \bar{c}_{n+i} x_{n+i}$$

$$= z^* - \sum_{k=1}^{n} \bar{c}_k x_k - \sum_{i=1}^{m} w_i^* \left(b_i - \sum_{j=1}^{n} a_{ij} x_j \right) \qquad \text{...(from (5.15) and (5.17))}$$

$$= \left(z^* - \sum_{i=1}^{m} w_i^* b_i \right) - \sum_{j=1}^{n} \left(\bar{c}_j - \sum_{i=1}^{m} a_{ij} w_i^* \right) x_j$$

$$= \left(z^* - \sum_{i=1}^{m} w_i^* b_i \right) - \sum_{j=1}^{n} \left(\sum_{i=1}^{m} a_{ij} w_i^* - c_j - \sum_{i=1}^{m} a_{ij} w_i^* \right) x_j$$

$$= \left(z^* - \sum_{i=1}^{m} w_i^* b_i \right) + \sum_{j=1}^{n} c_j x_j$$

And hence, $z^* - \sum_{i=1}^{m} w_i^* b_i = 0 \left(\text{since } z = \sum_{j=1}^{n} c_j x_j \right)$ ...(5.20)

Now, combining (5.19) and (5.20), we have

$$\sum_{j=1}^{n} c_j x_j^* = \sum_{i=1}^{m} w_i^* b_i$$

Now, by the optimality theorem (Theorem 5.3), it proves that w_i^* is the optimal solution for the dual problem. This establishes the equality of the objective functions at the optimal points.

Corollary 5.1: If either of the LP problems is unbounded, then the other problem must be infeasible.

Proof: This follows from the weak duality theorem. Suppose the primal (dual) problem has an unbounded solution, then the dual (primal) cannot have a feasible solution, since the dual (primal) objective function value will immediately provide an upper (lower) bound on the objective function value for the primal (dual) and then it will contradict our assumption. So, if the LP problem is unbounded, then the other problem must be infeasible.

Corollary 5.2: If either LP is infeasible, then the other LP is either infeasible or unbounded.

Proof: This follows from the strong duality theorem, which states that if either problem has a finite optimal solution, then so does the dual (primal) problem, and their objective function values are same at the optimal point. Thus, if one problem is infeasible, then the other cannot have an optimal solution, and so it must be either infeasible or unbounded.

5.7. Economic Interpretation of Duality

Consider the following primal problem in "n" economic activities and "m" resources.

$$\text{Maximize } z = \sum_{j=1}^{n} c_j x_j$$

Subject to the constraints

$$\sum_{j=1}^{n} a_{ij} x_j \leq b_i, i = 1, 2, \dots m$$

$$x_j \geq 0, j = 1, 2, \dots, n,$$

Where c_j is the revenue per unit of the activity j, b_i is the maximum availability of the resource i, and a_{ij} is the consumption of resource i per unit of activity j. Then the dual of the given problem is given by:

$$\text{Minimize } W = \sum_{i=1}^{m} b_i w_i$$

Subject to the constraints

$$\sum_{i=1}^{m} a_{ij} w_i \geq c_j, j = 1, 2, \dots, n, w_i \geq 0, i = 1, 2, \dots m$$

Interpretation of Dual Variables: We know that at the optimal points the value of the primal objective function equals the dual objective function. That is, $\sum_{j=1}^{n} c_j x_j^* = \sum_{i=1}^{m} b_i w_i^*$ ...(5.21)

The LHS of (5.21) gives the value of the maximal revenue of the primal problem, whereas the RHS is the sum-product of the availability of the ith resource and its worth per unit. Hence, w_i denotes the worth or imputed cost per unit of the ith resource.

Interpretation of Dual Constraints: At any feasible point for the primal, we have:

Objective coefficient of $x_j = \sum_{i=1}^{m} a_{ij} w_i - c_j; j = 1, 2, \dots, n;$

Since c_j is the revenue per unit of the activity j, $\sum_{i=1}^{m} a_{ij} w_i = \sum_{i=1}^{m}$ (consumption of resource i per unit of activity j) * (worth per unit of the ith resource) must represent imputed cost of all the resources to produce one unit of activity "j." Therefore, the quantity $\sum_{i=1}^{m} a_{ij} w_i - c_j$ represents the reduced cost of the activity "j." In the simplex calculation, with maximization as an objective function, it is economically advantageous to increase the level of activity "j" if the revenue of one unit of activity "j" > imputed cost per unit of activity "j."

Important Points

(i) The value of the objective function of the maximization problem for any feasible solution is always a lower bound for the value of the objective function of the minimization problem.

(ii) The value of the objective function of the minimization problem for any feasible solution is always an upper bound for the value of the objective function of the maximization problem.

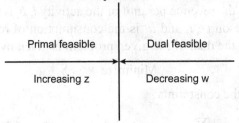

Fig. 5.3

(iii) If the primal (dual) has a feasible solution, but its value of the objective function is unbounded, then the dual (primal) problem has an infeasible solution.

(iv) If the primal (dual) problem has an infeasible solution, then the dual (primal) problem has an unbounded solution.

(v) Dual variables are called shadow prices of the limited resources, which are the imputed costs of respective resources.

(vi) If there are feasible solutions $x_j^* \geq 0 \ \forall \ j$ and $w_j^* \geq 0 \ \forall \ i$ for the symmetric primal and dual linear programs, respectively, and their value of the objective functions are same at these solutions, then these are the optimal solutions for their respective problems.

5.8. Dual Simplex Method

When we solve an LPP using the simplex method, we start with the feasible solution and try to achieve optimality. While doing this, we are actually trying to achieve the feasibility of our dual solution. In the primal simplex method, we move from one feasible basis to another until we get the optimal feasible basis, which is also a feasible basis for the dual problem. Thus, in solving the primal by the simplex method, the dual solution remains infeasible when the primal solution is feasible but nonoptimal. The primal problem can also be solved by starting with the dual feasible basis, where the optimality condition of the primal problem will be satisfied and we try to obtain a primal feasible basis by moving from one dual feasible basis to another. This is the approach used in the **dual simplex method.**

In the dual simplex method, we start with the solution which is infeasible (it does not satisfy the non-negative restriction) but optimal (it satisfies the optimality condition), and by moving from iteration to iteration we try to achieve its feasibility by maintaining its optimality. As soon as we get a feasible solution, we terminate the process.

This method sometimes is very helpful in solving problems without using artificial variables, and is also very useful in carrying out sensitivity analysis, in solving integer programming problems, and so on. The main difficulty in applying this method is to get the initial dual feasible basis, which sometimes is not easy to obtain. In the (primal) simplex method, while moving from one table to another we first select the entering non-basic variable, and then the departing variable from the basis is selected. Whereas, in the dual simplex Method, the leaving variable will be decided first and then followed by the entering variable. The rules for deciding the leaving and entering variables in the dual simplex algorithm are given as follows.

Feasibility condition: The basic variable having the most positive value is chosen as a leaving variable from the basis. When all the basic variables become non-negative, then the process terminates and the optimal as well as the feasible solution is reached.

Optimality condition: To decide the non-basic variable which will enter the basis, the following conditions are employed.

$$\text{(For maximization problem)} \frac{x_{Br}}{y_{rk}} = \text{Max} \left(\frac{z \text{ row coefficients}}{y_{rj}}, y_{rj} < 0 \right)$$

$$\text{(For maximization problem)} \frac{x_{Br}}{y_{rk}} = \text{Min} \left(\frac{z \text{ row coefficients}}{y_{rj}}, y_{rj} < 0 \right)$$

Here, the index r is used for the $i = r^{th}$ row basic variable leaving from the basis, and the k index is used for the $j = k^{th}$ column variable, which is a non-basic variable entering the basis.

If no $y_{rj} < 0$, we will terminate the process and there is no feasible solution to the problem.

The working procedure of the simplex method is different from the dual simplex method in the following ways:

(i) In the simplex method, we start with the feasible but non-optimal basis, whereas in the dual simplex method, we start with the dual feasible basis.

(ii) In the simplex method, we first choose the entering variable and then the departing or leaving variable from the basis, whereas in the dual simplex method, it is the reverse.

(iii) In the simplex method, by moving from one feasible basis to another, we try to achieve the optimal solution, whereas in the dual simplex method, by moving from one optimal basis to another, we try to achieve the feasible solution.

Example 5.12. *Use the dual simplex method to solve the following LPP:*

$$Minimize \ z = 6x_1 + 8x_2$$

Subject to

$$x_1 + x_2 \geq 30$$
$$5x_1 + 2x_2 \geq 60$$
$$x_1 + x_2 \geq 0$$

Solution: Multiply both constraints by -1 to convert the $\geq$ to $\leq$ and re-write the resulting problem as:

$$-x_1 - x_2 + x_3 = -30$$
$$-5x_1 - 2x_2 + x_4 = -60$$
$$z - 6x_1 - 8x_2 = 0$$
$$x_1, x_2, x_3, x_4 \geq 0$$

The initial dual simplex table of the previous problem is:

Basic	x_1	x_2	x_3	x_4	Solution
x_3	-1	-1	1	0	-30
x_4	-5	-2	0	1	-60
z	-6	-8	0	0	0

The optimality condition of the current table is satisfied, as all elements in the z-row are ≤ 0. At the same time it is also infeasible, as the value of both the basic variables is ≤ 0. As per algorithm discussed previously, x_4 will leave the basis, as it has the most negative value. The following table shows how the optimality condition is used to decide the entering variable.

	$j=1$	$j=1$
Non – basis	x_1	x_2
z-row value (z_j)	-6	-8
$y_{4j}(<0)$	-5	-2
Ratio z_j/y_{4j}	1.2	4

So, x_1 enters the basis (since it has the minimum ratio). The next table is obtained using the normal row operation.

Basic	x_1	x_2	x_3	x_4	Solution
x_3	0	-0.6	1	-0.2	-18
x_1	1	0.4	0	-0.2	12
z	0	-5.6	0	-1.2	72
Ratio		9.33		6	

x_3 will leave the basis and x_4 will re-enter. The resulting table is given as follows:

Basic	x_1	x_2	x_3	x_4	Solution
x_4	0	3	-5	1	90
x_1	1	1	-1	0	30
z	0	-2	-6	0	180

Now, both the optimality and feasibility conditions are satisfied and the previous table is the optimal table of the given problem. It can be clearly seen that in all iterations, optimality is maintained; that is the elements in the z-row in all tables are ≤ 0, and as we move from one table to another, the solution also moves toward the feasibility. Hence, $x_1 = 30$ and $x_2 = 0$ is the optimal basis feasible solution, and the corresponding minimum value of the objective function is 180.

Exercises

1. Explain the key relationships between primal and dual problems.

2. Write a short note on duality in linear programming.

3. Prove that the dual of the dual of a given primal is again primal.

4. State and prove:

   (*i*) Weak Duality Theorem

   (*ii*) Strong Duality Theorem

   (*iii*) Complementary Slackness Theorem

5. Write the dual associated with each of the following linear programming problems:

   (*i*) Minimize $z = 10x_1 + 2x_2$

   Subject to $x_1 + x_2 \geq -1$

   $$x_1 + x_2 \geq -2$$

   $$2x_1 - x_2 \geq -1$$

   $$x_1, x_2 \geq 0$$

   (*ii*) Maximize $Z = 2x_1 + 3x_2$

   Subject to $2x_1 - 2x_2 \geq 4$

   $x_1 + 0.25x_2 \leq 1$, x_1 and x_2 are unrestricted in sign

   (*iii*) Minimize $Z = 7x_1 + 5x_2$

   Subject to $x_1 + x_2 = 1$

   $$-2x_1 + x_2 \leq 3$$

   $$-x_1 + x_2 \geq 1$$

   $x_1 \geq 0$, x_2, x_3 are unrestricted in sign

6. Use duality theory to determine whether $x_1 = 0$, $x_2 = 6.111$, $x_3 = 6.111$, and $x_4 = 3.333$ is an optimal solution of the following linear programming problem:

   $$\text{Max } z = x_1 + 2x_2 + 1.3x_3 - 0.5x_4$$

   Subject to the constraint

$$2.5x_1 + 2x_2 + x_3 - x_4 \le 15$$
$$2x_1 + x_2 + 2x_3 + 0.5x_4 \le 20$$
$$1.5x_1 + 2x_2 + x_3 + 2x_4 \le 25$$

and
$$x_1, x_2, x_3, x_4 \ge 0$$

What if the x_1 coefficient in the objective function is changed to 1.5 and the constraints remain the same?

7. Give the economic interpretation of duality.

8. Write the *dual* of the following linear programming problem and graphically determine the optimal values of the dual variables:
$$\text{Max } z = 3x_1 + 2x_2$$

Subject to

$$2x_1 + x_2 \le 25$$
$$x_1 + x_2 \le 15$$
$$x_1 \le 8$$

and
$$x_1, x_2 \ge 0$$

9. The following are the initial and optimal table of a given linear programming problem with minimization as an objective function.

Basic	x_1	x_2	x_3	x_4	x_5	x_6	x_7	Solution
x_5	1	2	-1	2	1	0	0	b_1
x_6	1	-2	0	1	0	1	0	b_2
x_7	0	0	1	0	0	0	1	b_3
z	-1	0	0	1	0	0	0	0

(Initial Table)

Basic	x_1	x_2	x_3	x_4	x_5	x_6	x_7	Solution
x_4	0.6667	0	0	1	0.3333	0.3333	0.3333	20
x_3	0	0	1	1	0	0	1	42
x_2	f	1	0	0	0.16667	-0.3333	0.16667	7
z	a	b	c	d	-0.3333	e	-0.3333	20

(Optimal Table)

(i) Identify the optimal basis inverse

(ii) Find the value , and b_1, b_2 and b_3

(iii) Obtain the dual solution

(iv) Determine the values of a, b, c, d, e, and f.

10. Explain:
   (i) Weak Duality

   (ii) Strong Duality

   (iii) Complementary Slackness Property in Duality.

11. Consider the following LPP:
$$\text{Max } z = 15x_1 + 12\,x_2 + 8x_3$$

Subject to
$$2x_1 + x_2 + x_3 \leq 50$$
$$2x_1 + x_2 + 2x_3 \leq 65$$
$$x_1 + 2x_2 - x_3 \leq 45$$

and
$$x_1, x_2, x_3 \geq 0$$

Check the optimality and feasibility conditions by constructing the entire simplex tableau associated with each of the following basic solutions:

(i) Basic variable $= (x_1, s_2, x_2)$, $= \begin{pmatrix} 0.6667 & 0 & -0.3333 \\ -1 & 1 & 0 \\ -0.3333 & 0 & 0.6667 \end{pmatrix}$

(ii) Basic variable $= (x_1, x_3, x_2)$, $= \begin{pmatrix} 1.6667 & -1 & -0.3333 \\ -1 & 1 & 0 \\ -1.3333 & 0 & 0.6667 \end{pmatrix}$.

12. Explain the dual simplex method used for solving a given LPP. How does it differ from the normal simplex method?

13. Use the dual simplex method to solve the following LPP.
   (i) Max $z = -1.5x_1 - 4x_2$

   Subject to
$$x_1 + 2x_2 \geq 22$$

$$2x_1 + 3x_2 \leq 45$$
$$x_1 \leq 12$$

and
$$x_1, x_2 \geq 0$$

(*ii*)   Min $z = x_1 + 1.2x_2 + 0.5x_3$

Subject to

$$x_1 + 2x_2 + x_3 \geq 20$$
$$2x_1 + 2x_2 + x_3 \geq 35$$
$$x_1 + x_2 \geq 10$$

and
$$x_1, x_2, x_3 \geq 0$$

(*iii*)   Min $z = 2x_1 + 2x_2$

Subject to

$$x_1 + 2x_2 \geq 1.5$$
$$-2x_1 + x_2 \leq -1$$
$$2x_1 + 4x_2 \geq 1.8$$

and
$$x_1, x_2, x_3 \geq 0$$

CHAPTER 6

SENSITIVITY ANALYSIS

6.1. Introduction

In all linear programming problems, we make an assumption that the parameters a_{ij}, b_i, and c_j are certainly known to us, but in real life it is not true. The values of these parameters are always estimated based on experience and with the help of past data, which never remains the same and keeps changing with time. Sensitivity analysis is used to determine what changes will occur in the current optimal basic feasible solution corresponding to the changes in one or more parameter values. With the help of sensitivity analysis, we can determine the sensitivity of the parameters, that is, how sensitive the parameters are. There are some parameters which are very sensitive in that if we slightly change the value of the parameter, the solution changes; on the other hand, the solution due to the change in some parameters remains unchanged in some range of parameter values but beyond this range, there is a change in the solution. With the help of this analysis, we can determine the range of parameter values within which the current solution remains feasible as well optimal but obviously with the changed values of the variables.

By using the concepts of the construction of a simplex table and duality, which have been discussed earlier, we can determine the change in the final optimal basic feasible solution due to the change in parameter values, without solving the whole problem again from the beginning. There may be two types of changes in the optimal basic feasible solution: either the feasibility or the optimality of the current optimal basic feasible solution may get disturbed.

The feasibility of the optimal solution may get disturbed due to the changes in the following parameter values:

(i) Availability parameter b_i for the ith resource, $i = 1, 2$ .....m

(ii) Addition of a new constraint

(iii) Deletion of a variable

In these cases, the dual simplex method may be applied to regain the feasibility of the solution.

The optimality condition of the optimal solution may get disturbed due to the changes in the following parameter values:

(i) The cost coefficient c_j in the objective function, $j = 1, 2$ .....n

(ii) Addition of a new variable

(iii) Changes in the coefficients a_{ij} of a variable x_j in constraints $j = 1, 2$ .....n

(iv) Deletion of a constraint

Here the optimality of the current solution may be regained by applying the simplex method.

Note: (1) In case of any change in the parameter values of the given problem, the initial basic variables comprising slack or artificial variables and their coefficients remain the same, and hence corresponding to these initial basic variables, the basis inverse B–1 and the values of the dual variables in the final table also don't change. Therefore, with the help of these values, the remaining values in the final simplex table can be obtained using the relations given as follows:

(a) z-row coefficients = (values of dual variables) * (original changed vector)- (changed price corresponding to these vectors)

(b) Any other vector in the Simplex table = (basis inverse) * (changed vector if any)

(c) Changed values of basic variables = (basis inverse) * (changed original RHS vector)

(d) Changed value of the objective function = (values of dual variables) * (changed RHS vector)

Here instead of taking the complete changed vector, we can find the incremental changes in the final table as there are incremental changes in the initial table by ignoring all terms where there is no change in the initial table. With the help of this procedure, we can find the range of values of given parameters in which our current optimal basic feasible solution will remain feasible as well as optimal.

After applying the previously mentioned operations, it might happen that the final matrix is not in an appropriate form as is required, like the vectors corresponding to basic variables do not form identity matrix (this may happen if the coefficients corresponding to basic variables appearing in final table, changes in the original problem); in this case by applying the Gaussian elimination method, we can convert it to the desired form. Now in the final table, if the optimality or the feasibility condition has been disturbed, then after applying the simplex or dual simplex method, respectively, the optimality and the feasibility of the final solution may be regained.

Now let us consider the changes in different parameter values of any given linear programming problem and see how we can obtain the solution of the changed problem without solving the whole problem again.

6.2. Changes in the RHS Vector b_i

Here we assume that one or more b_i ($i = 1 \ldots m$) parameter values change in the model. As we have discussed earlier, due to this change there will be no change in the final simplex table except the RHS column or the values of the basic variable and hence the feasibility of the solution.

The new values of the basic variables appearing in the final simplex table can be obtained as follows:

(New values of basic variables in final simplex table) = (basis inverse in final simplex table) * (newly changed vector b in the original model)

$$x_B^* = B^{-1}b^* \text{ where } b^* \text{ is changed } b \text{ vector}$$

And the change in the value of the objective function corresponding to this change is (New value of obj. function) = (values of dual variables)* (changed vector b)

Example 6.1. *Consider the following LPP:*

Max $\qquad z = x_1 + 3x_2$

Subject to

$$x_1 \leq 5, \ x_1 + 2x_2 \leq 10, \ x_2 \leq 4 \ x_1, \ x_2 \geq 0.$$

Solution: Writing the given problem into its canonical form, we have:

Max z

Subject to

$$x_1 + s_1 = 5$$
$$x_1 + 2x_2 + s_2 = 10$$
$$x_2 + s_3 = 4$$
$$z - x_1 - 3x_2 = 0$$
$$x_1, x_2, s_1, s_2, s_3 \geq 0$$

Initial Table

Basic	x_1	x_2	s_1	s_2	s_3	Solution	Ratio
s_1	1	0	1	0	0	5	
s_2	1	2	0	1	0	10	5
s_3	0	1	0	0	1	4	4
z	−1	−3	0	0	0	0	

x_2 enters and s_3 departs.

First Iteration

Basic	x_1	x_2	s_1	s_2	s_3	Solution	Ratio
s_1	1	0	1	0	0	5	5
s_2	1	0	0	1	−2	2	2
x_2	0	1	0	0	1	4	
z	−1	0	0	0	3	12	

x_1 enters and s_2 departs.

Final Table

Basic	x_1	x_2	s_1	s_2	s_3	Solution
s_1	0	0	1	−1	2	3
x_1	1	0	0	1	−2	2
x_2	0	1	0	0	1	4
z	0	0	0	1	1	14

Since all the elements in the z row are ≥ 0, the optimality condition is now satisfied. Here, the optimal solution is $x_1 = 2$, $x_2 = 4$, *and* $z = 14$.

Now we change the RHS vector to $b^* = \begin{pmatrix} 12 \\ 15 \\ 8 \end{pmatrix}$. We have the following information from the table.

$$Basis\ Inverse = B^{-1} = \begin{pmatrix} 1 & -1 & 2 \\ 0 & 1 & -2 \\ 0 & 0 & 1 \end{pmatrix}$$ and values of dual variables or shadow prices which represent the worth of all resources are 0, 1, 1.

Hence, the values of the new basic variables are $B^{-1}b^* = (11, -1, 8)$ and $z^* = 23$

Here $b^* = \begin{pmatrix} 10 \\ 15 \\ 8 \end{pmatrix} = b + \Delta b = \begin{pmatrix} 5 \\ 10 \\ 4 \end{pmatrix} + \begin{pmatrix} 5 \\ 5 \\ 4 \end{pmatrix} where\ \Delta b = \begin{pmatrix} 5 \\ 5 \\ 4 \end{pmatrix}$

By incremental analysis we can also obtain the changed values of the basic variables as:

Old values of basic variables in the final table + incremental values

$$\left(B^{-1}\Delta b\right) = \begin{pmatrix} 3 \\ 2 \\ 4 \end{pmatrix} + \begin{pmatrix} 1 & -1 & 2 \\ 0 & 1 & -2 \\ 0 & 0 & 1 \end{pmatrix} * \begin{pmatrix} 5 \\ 5 \\ 4 \end{pmatrix} = \begin{pmatrix} 3 \\ 2 \\ 4 \end{pmatrix} + \begin{pmatrix} 8 \\ -3 \\ 4 \end{pmatrix} = \begin{pmatrix} 11 \\ -1 \\ 8 \end{pmatrix}$$

Similarly, the new value of the objective function = the old value of the objective function + the incremental value of the objective function = 14 +

(values of dual basic variables) $* \left(\Delta b\right) = 14 + \begin{pmatrix} 0 & 1 & 1 \end{pmatrix} * \begin{pmatrix} 5 \\ 5 \\ 4 \end{pmatrix} = 23$.

Now the RHS vector of the final table has become infeasible (*i.e.*, the values of the basic variables have become negative), so we will apply the dual simplex method to get the feasible as well as the optimal solution. Changing the solution column in the final table of the original problem by 11, −1, 8, and 23, we have:

Basic	x_1	x_2	s_1	s_2	s_3	Solution
s_1	0	0	1	-1	2	11
x_1	1	0	0	1	-2	-1
x_2	0	1	0	0	1	8
z	0	0	0	1	1	23
Ratio					-0.5	

Clearly, x_1 will leave the basis, and since -2 is the only negative value in the row corresponding to the leaving variable, s_3 enters the basis in the next table.

Basic	x_1	x_2	s_1	s_2	s_3	Solution
s_1	1	0	1	0	0	10
s_3	-0.5	0	0	-0.5	1	0.5
x_2	0.5	1	0	0.5	0	7.5
z	0.5	0	0	1.5	0	22.5

Since both the optimality and feasibility conditions are now satisfied, the optimal solution has been reached. Hence, $x_1 = 0$ and $x_2 = 7.5$ is the optimal solution of the changed problem, and the corresponding value of z is 22.5.

The graphical solution of the original and modified problem is shown as follows.

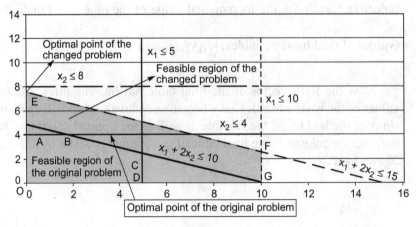

Fig. 6.1

In the previous graph, the solid lines denote the original constraints, the region OABCD gives the feasible region of the original problem, and the optimal solution of this problem occurs at the extreme point B (2, 4) with the corresponding value of z as 14. The dotted lines denote the changed constraints when the b is changed to b^*, and the region surrounded by OEFG gives the feasible region of the changed problem. Here, the optimum solution occurs at the extreme point E (0, 7.5), and the corresponding value of the objective function is 22.5. Hence, there is an increment of 8.5 in the value of the objective function because of the new b vector. It can also be seen that the constraint $x_2 \le 8$ now becomes redundant.

6.2.1. Range of b_i's

We can find the range of each $b_i (i = 1, ...m)$ parameter in which it can change so that the current solution remains feasible as well as optimal.

Range for b_1: We have $b^* = b + \Delta b = \begin{pmatrix} b_1 \\ b_2 \\ b_3 \end{pmatrix} + \begin{pmatrix} \Delta b_1 \\ 0 \\ 0 \end{pmatrix}$; that is, we assume

that there is a change in the b_1 component of the b vector. Due to this change, we want our solution to remain feasible as well as optimal, that is,

$$B^{-1}b^* = B^{-1}(b + \Delta b) B^{-1}b + B^{-1} \begin{pmatrix} \Delta b_1 \\ 0 \\ 0 \end{pmatrix}$$

$= (old\ solution) + (incremental\ change\ in\ the\ solution) \ge 0$

$$\Rightarrow \begin{pmatrix} 3 \\ 2 \\ 4 \end{pmatrix} + \begin{pmatrix} 1 & -1 & 2 \\ 0 & 1 & -2 \\ 0 & 0 & 1 \end{pmatrix} * \begin{pmatrix} \Delta b_1 \\ 0 \\ 0 \end{pmatrix} \ge 0$$

$\Rightarrow \qquad\qquad 3 + \Delta b_1 \ge 0$

$\Rightarrow \qquad\qquad \Delta b_1 \ge -3$

$\Rightarrow \qquad\qquad -3 \le \Delta b_1 \le \infty$

$\Rightarrow \qquad\qquad 2 \le b_1^* \le \infty$

Range for b_2: We have $b^* = b + \Delta b = \begin{pmatrix} b_1 \\ b_2 \\ b_3 \end{pmatrix} + \begin{pmatrix} 0 \\ \Delta b_2 \\ 0 \end{pmatrix}$; that is, we assume that

there is a change in the b_2 component only of the b vector. Due to this change, we want our solution to remain feasible as well as optimal; that is:

$$B^{-1}b^* = B^{-1}(b+\Delta b)\, B^{-1}b + B^{-1}\begin{pmatrix} 0 \\ \Delta b_2 \\ 0 \end{pmatrix} \geq 0$$

$\Rightarrow$ (old solution) + (incremental change in the solution due to the change in b vector) ≥ 0

$$\Rightarrow \quad \begin{pmatrix} 3 \\ 2 \\ 4 \end{pmatrix} + \begin{pmatrix} 1 & -1 & 2 \\ 0 & 1 & -2 \\ 0 & 0 & 1 \end{pmatrix} * \begin{pmatrix} 0 \\ \Delta b_2 \\ 0 \end{pmatrix} \geq 0$$

$\Rightarrow \qquad\qquad\qquad 3 - \Delta b_2 \geq 0 \quad$ and $\quad 2 + \Delta b_2 \geq 0$

$\Rightarrow \qquad\qquad\qquad\quad \Delta b_2 \leq 3 \quad$ and $\quad \Delta b_2 \geq -2$

$\Rightarrow \qquad\qquad\qquad -\infty \leq \Delta b_2 \leq 3 \quad$ and $\quad -2 \leq \Delta b_2 \leq \infty$

$\Rightarrow$ By combining these two inequalities we get

$$-2 \leq \Delta b_2 \leq 3 \quad \text{and} \quad 8 \leq b_2^* \leq 13.$$

Range for b_3: Now, we have $b^* = b + \Delta b = \begin{pmatrix} b_1 \\ b_2 \\ b_3 \end{pmatrix} + \begin{pmatrix} 0 \\ 0 \\ \Delta b_3 \end{pmatrix}$; that is, we as-

sume that there is a change in the b_3 component only of the b vector. Due to this change, we want our solution to remain feasible as well as optimal, that

is: $B^{-1}b^* = B^{-1}(b + \Delta b) = B^{-1}b + B^{-1}\begin{pmatrix} 0 \\ 0 \\ \Delta b_3 \end{pmatrix}$

= (*old solution*) (*incremental change in the solution due to the change in the b vector*) ≥ 0)

$$\Rightarrow \quad \begin{pmatrix} 3 \\ 2 \\ 4 \end{pmatrix} + \begin{pmatrix} 1 & -1 & 2 \\ 0 & 1 & -2 \\ 0 & 0 & 1 \end{pmatrix} * \begin{pmatrix} 0 \\ 0 \\ \Delta b_3 \end{pmatrix} \geq 0$$

$\Rightarrow \qquad\qquad 3 + 2\Delta b_3 \geq 0; 2 - 2\Delta b_3 \geq 0 \quad$ and $\quad 4 + \Delta b_3 \geq 0$

$\Rightarrow \qquad\qquad \Delta b_3 \geq -\left(\dfrac{2}{3}\right); \Delta b_3 \leq 1 \quad$ and $\quad -4 \leq \Delta b_3$

$\Rightarrow$ By combining all three inequalities we get

$$-\left(\dfrac{3}{2}\right) \leq \Delta b_3 \leq 1 \quad \text{and} \quad \text{hence} \quad \left(\dfrac{5}{2}\right) \leq b_3^* \leq 5$$

Hence, these are the ranges in which if any component of the b (RHS) vector of constraints changes, then the current solution will remain feasible as well as optimal, but the values of the current basic variables may change and the values of the dual variables or shadow prices can be used to find the change in the objective function. If there is a change in any of the components of the b vector outside the given range, then the basis and the corresponding basic feasible solution will get changed. These ranges work well only when there is a change in one component of the b vector and the other components remain the same.

The following is the sensitivity report of the same problem obtained using WinQSB-(see **Appendix**):

08-21-2016 10:58:47	Constraint	Direction	Shadow Price	Right Hand Side	Allowable Min. RHS	Allowable Max. RHS
1	C1	<=	0	5.0000	2.0000	M
2	C2	<=	1.0000	10.0000	8.0000	13.0000
3	C3	<=	1.0000	4.0000	2.5000	5.0000

Fig. 6.2

*M is a very large positive number.

Interpretations of the previous output:

- The allowable increase for b1 is infinity and can be decreased at most by 3; that is, $-3 \leq \Delta b1 \leq \infty$ and hence $2 \leq b_1^* \leq \infty$.

- b_2 can be increased at the most by 3 and can be decreased up to 2; that is, $-2 \leq \Delta b_2 \leq 3$ and hence $8 \leq b_2^* \leq \infty$.

- The allowable increase for b_3 is up to 5 and can be decreased at most by 1.5; that is, $-15 \leq \Delta b_3 \leq 1$ and hence $2.5 \leq b_3^* \leq 5$.

6.2.2 Simultaneous Changes in $b_i's$

Now we are interested in finding the range of $b_i' s$ when several of them change simultaneously. There is a 100% rule for finding these ranges which calculates, the actual change in each b_i as a percentage of allowable change in that component of the b vector as determined previously and then adds these percentages. If the total percent change remains within 100%, then the current solution will remain feasible and the values of the dual variables still can be used to find the change in the objective function; otherwise, the present solution may become infeasible and we will be required to find an alternative optimum solution.

In the Example (6.1), suppose we have changed all the components simultaneously, so if we calculate percent change, we get:

$$\text{Allowable \% change in } b_1 = \frac{\text{Actual change in } b_1}{\text{Maximum Allowable change in } b_1} \times 100$$

$$= \frac{5}{\infty} \times 100 = 0$$

Similarly,

$$\text{Allowable \% change in } b_2 = \frac{\text{Actual change in } b_2}{\text{Maximum Allowable change in } b_2} \times 100$$

$$= \frac{5}{3} \times 100 = 166.667$$

$$\text{Allowable \% change in } b_3 = \frac{\text{Actual change in } b_3}{\text{Maximum Allowable change in } b_3} \times 100$$

$$= \frac{4}{1} \times 100 = 400$$

(Here, if there is a decrease in b_i, we will take the maximum allowable decrease, and in case of increase in the component value, we will take maximum allowable increase.)

Now if we sum up these percentages, it is more than 100%, and the solution becomes infeasible as seen from the previous calculation. It should also be noted that a percentage sum more than 100% does not mean the solution will always become infeasible, whereas if the percentage sum is less than 100%, then the current solution will always remain feasible as shown by the following example:

In the Example (6.1), let $b^* = \begin{pmatrix} 10 \\ 11 \\ 4.5 \end{pmatrix}$. Then

$$\text{allowable \% change in } b_1 = \frac{\text{Actual change in } b_1}{\text{Maximum Allowable change in } b_1} \times 100$$

$$= \frac{5}{\infty} \times 100 = 0$$

Similarly,

$$\text{allowable \% change in } b_2 = \frac{\text{Actual change in } b_2}{\text{Maximum allowable change in } b_2} \times 100$$

$$= \frac{1}{3} \times 100 = 33.33$$

$$\text{Allowable \% change in } b_3 = \frac{\text{Actual change in } b_3}{\text{Maximum allowable change in } b_3} \times 100$$

$$= \frac{0.5}{1} \times 100 = 50$$

Now the sum of these percentages is 83.33, and it's less than 100%, and hence the present solution should remain feasible. And the new values of the optimal basic variables are is given by:

$$B^{-1}b^* = \begin{pmatrix} 1 & -1 & 2 \\ 0 & 1 & -2 \\ 0 & 0 & 1 \end{pmatrix} \begin{pmatrix} 10 \\ 11 \\ 4.5 \end{pmatrix} = \begin{pmatrix} 8 \\ 2 \\ 4.5 \end{pmatrix}$$ and the corresponding maximum value

of the objective function is $z^* = 15.5$.

6.3. Addition of a New Constraint

Sometimes after getting the solution of the problem, we need to add a new constraint which might be due to additional constraints on present resources, the addition of some new resources, the introduction of some additional technical constraints, or not having considered an issue earlier due to some reason, and so on. When a new constraint is added to the problem, it never improves the value of the objective function; either it remains the same since the constraint is redundant, or it deteriorates the value of the objective function since it constricts the region of the feasible solution and hence the value of the objective function.

Hence, in order to check the effect of the new constraint on the current optimal feasible solution, we will put the current solution in the new constraint to check if the constraint is satisfied. If yes, then the current optimal solution will remain the same for the changed problem; the new constraint is redundant and is not playing any role in finding the solution of the problem. But if the constraint gets violated by the present optimal solution, then we introduce the new constraint in the last row of the final simplex table by expressing the whole constraint in terms of non-basic variables and taking the slack variable of the constraint as a basic variable in the table. Here, in this case, the feasibility condition of the current solution is not satisfied, and we apply the dual simplex method to regain it again by maintaining the optimality condition. In this way, we will get the new optimal basic feasible solution.

6.3.1. When the Current Solution Satisfies the New Constraint

To illustrate the above mentioned procedure, consider the Example (6.1) again.

The solution of the given problem is $x_1 = 2$, $x_2 = 4$, *and* $z = 14$. In the same problem, if we add the following new constraint

$$3x_1 + x_2 \le 15$$

then we see that for the current solution, the constraint remain satisfied. Hence, the solution remains the same for the new problem, and the new constraint becomes redundant. The situation can be easily visualized through the following graph:

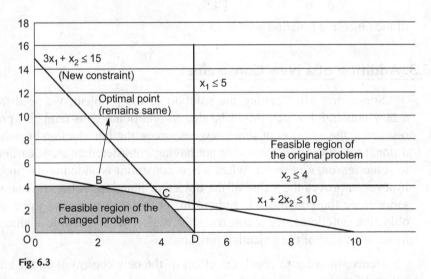

Fig. 6.3

6.3.2. When the Current Solution Fails to Satisfy the New Constraint

Let $3x_1 + x_2 \le 8$ be the new constraint in the same Example (6.1) considered previously. Clearly the constraint gets violated by the current solution, and hence we need to find the new solution. From the final simplex table of the given problem, we obtain the following:

$$x_1 + s_2 - 2s_3 = 2 \quad \Rightarrow \quad x_1 = 2 - s_2 + 2s_3$$

Similarly, $\qquad\qquad x_2 + s_3 = 4 \quad \Rightarrow \quad x_2 = 4 - s_3$

Substituting the expression of x_1 and x_2 and adding a slack variable in the given constraint, we get:

$$3(2 - s_2 + 2s_3) + (4 - s_3) + s_4 = 8$$

Here s_4 is a new slack variable added in this constraint. Now on simplification, we get:

$$-3s_2 + 5s_3 + s_4 = -2$$

On adding this new constraint in the final simplex table and **applying the dual simplex method** we have:

Basic	x_1	x_2	s_1	s_2	s_3	s_4	Solution
s_1	0	0	1	−1	2	0	3
x_1	1	0	0	1	−2	0	2
x_2	0	1	0	0	1	0	4
s_4	0	0	0	−3	5	1	−2
z	0	0	0	1	1	0	14
Ratio				− 0.3333			

s_4 leaves and s_2 enters the basis.

Basic	x_1	x_2	s_1	s_2	s_3	s_4	Solution
s_1	0	0	1	0	0.3333	-0.3333	3.6667
x_1	1	0	0	0	-0.3333	0.3333	1.3333
x_2	0	1	0	0	1	0	4
s_2	0	0	0	1	−1.6667	−0.3333	0.6667
z	0	0	0	0	2.6667	0.3333	13.3333

Since both the optimality and feasibility conditions are satisfied, our new optimal solution is $x_1 = 1.33$, $x_2 = 4$, and $z = 13.33$, and the same is shown graphically as follows:

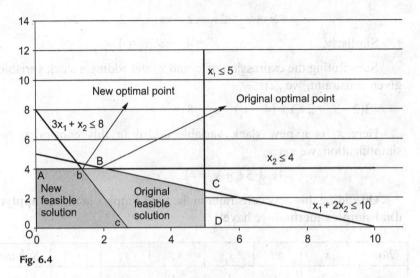

Fig. 6.4

Further, if we add the constraint $x_2 \geq 1$ in the new obtained solution, then the new solution will remain unchanged and the newly added constraint is ineffective; this can be seen in the following graph:

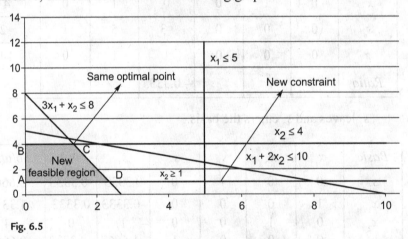

Fig. 6.5

But if we change the constraint to $x_2 \leq 3$, which is not satisfied by our new obtained solution, the solution will change again because of this constraint, and it can be easily seen from the following tables and graphs.

From the final dual simplex table of the changed problem, we obtain the following:

$$x_2 + s_3 = 4 \quad \Rightarrow \quad x_2 = 4 - s_3$$

Substituting the value x_2 and adding the slack variable s_5 in this new constraint, we get:

$$(4 - s_3) + s_5 = 3 \quad \Rightarrow \quad -s_3 + s_5 = -1$$

On adding this new constraint in the final dual simplex table of the changed problem, we have:

Basic	x_1	x_2	s_1	s_2	s_3	s_4	s_5	Solution
s_1	0	0	1	0	0.3333	−0.3333	0	3.6667
x_1	1	0	0	0	− 0.3333	0.3333	0	1.3333
x_2	0	1	0	0	1	0	0	4
s_2	0	0	0	− 3	5	1	0	0.6667
s_5	0	0	0	0	−1	0	1	−1
z	0	0	0	1	1	0	0	13.3333
Ratio					− 0.333			

s_5 leaves and s_3 enters

Basic	x_1	x_2	s_1	s_2	s_3	s_4	s_5	Solution
s_1	0	0	1	0	0	− 0.3333	0.3333	3.3334
x_1	1	0	0	0	0	0.3333	− 0.3333	1.6666
x_2	0	1	0	0	0	0	1	3
s_2	0	0	0	1	0	− 0.3333	−1.6667	2.3334
s_3	0	0	0	0	1	0	−1	1
z	0	0	0	0	0	0.3333	2.6667	10.6666

Since all the entries in the z-row are ≥ 0, the optimum solution has been reached. Our new optimum solution is $x_1 = 1.666$, $x_2 = 3$, and the corresponding maximum value of z is 10.666. The same can be seen from the following graph:

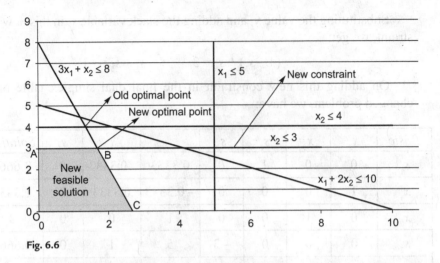

Fig. 6.6

Now our optimal solution has been changed to $x_1 = 1.67$, $x_2 = 3$, $z = 10.67$. Hence, it can be noted that on adding each new constraint, if the constraint is active and changes the solution, then every time value of the objective function deteriorates.

6.4. Adding a New Activity or a Variable

When we add a new activity or a variable in the given problem, the cost coefficient of that variable will appear in the objective function and a vector corresponding to that variable will be added in the constraints, which will give the coefficient of that variable in each constraint. The coefficient of the new variable (activity) in the ith constraint represents the consumption of the ith resource by one unit of the new activity. Now we have to see the effect of the addition of this activity on the current optimal solution. As we discussed earlier, it may affect the optimality condition of the current solution and the feasibility condition will not be affected by it at all. To check the optimality condition, we calculate the z coefficient corresponding to this variable in the final simplex table. If the optimality condition still remains satisfied, then there is no change in the current optimal solution and the new activity is not profitable to produce. On the other hand, if the optimality condition gets disturbed, then by applying the simplex method, we try to obtain a new optimum solution which satisfies the optimality as well as the feasibility condition. The newly obtained solution will contain the newly introduced variable, and hence it will be profitable to do that activity.

If we consider the same Example (6.1) where the solution is $x_1 = 2, x_2 = 4$, and $z = 14$ and add a new variable x_3 in the problem with $c_3 = 4$ and vector

$a_3 = \begin{pmatrix} 1 \\ 2 \\ 4 \end{pmatrix}$, then the column corresponding to this variable in the final table can

be obtained as:

$$B^{-1}a_3 = \begin{pmatrix} 1 & -1 & 2 \\ 0 & 1 & -2 \\ 0 & 0 & 1 \end{pmatrix}\begin{pmatrix} 1 \\ 2 \\ 4 \end{pmatrix} = \begin{pmatrix} 7 \\ -6 \\ 4 \end{pmatrix}$$

And the z coefficient corresponding to the new variable = (values of dual variables) * $(a_3$ vector$) - c_3$

$$= \begin{pmatrix} 0 & 1 & 1 \end{pmatrix}\begin{pmatrix} 1 \\ 2 \\ 4 \end{pmatrix} - 4 = 2$$

In this case, the optimality condition remains satisfied, and hence there is no change in the current optimal solution. But if we change the vector

$a_3 = \begin{pmatrix} 1 \\ 2 \\ 4 \end{pmatrix}$ to $a_3 = \begin{pmatrix} 1 \\ 1 \\ 1 \end{pmatrix}$, then we

Have

$$B^{-1}a_3 = \begin{pmatrix} 1 & -1 & 2 \\ 0 & 1 & -2 \\ 0 & 0 & 1 \end{pmatrix}\begin{pmatrix} 1 \\ 1 \\ 1 \end{pmatrix} = \begin{pmatrix} 2 \\ -1 \\ 1 \end{pmatrix} \text{ and }$$

$$z \text{ coefficient} = \begin{pmatrix} 0 & 1 & 1 \end{pmatrix}\begin{pmatrix} 1 \\ 1 \\ 1 \end{pmatrix} - 4 = -2$$

Now the optimality condition is violated, and hence if we add these values in the final simplex table and apply the simplex method, we get the following solution:

Basic	x_1	x_2		s_1	s_2	s_3	Solution	Ratio
s_1	0	0	2	1	− 1	2	3	1.5

x_1	1	0	– 1	0	1	– 2	2	
x_2	0	1	1	0	0	1	4	4
z	0	0	–2	0	1	1	14	

x_3 enters and s_1

Basic	x_1	x_2	x_3	s_1	s_2	s_3	Solution
x_3	0	0	1	0.5	–0.5	1	1.5
x_1	1	0	0	0.5	0.5	– 1	3.5
x_2	0	1	0	– 0.5	0.5	0	2.5
z	0	0	0	1	0	3	17

We can clearly see that the new value of z is 17 and is more than the original value of z (14). Hence, it is profitable to produce the new activity, since it increases the value of the objective function by 3 units.

6.5. Changes in the Objective Function Coefficients

6.5.1. Changes in the Objective Function Coefficients of Non-Basic Variables

If there is a change in cost coefficient c_j of the variable x_j, which is a non-basic variable in the final table, then the only change in the final simplex table corresponding to this change will be in the z-row coefficient of this variable. The new value of the z-row coefficient can be determined by finding (LHS-RHS) of the dual constraint corresponding to variable x_j for the present values of the dual variables:

z-row coefficient of x_j = (value of dual variables) * (coefficient vector a_j) – changed value of c_j

If the optimality condition is still satisfied, then the current solution will remain feasible as well as optimal; otherwise, we allow x_j to enter the basis and apply the simplex method to find the new optimal solution.

Range of change of cj: To find the range of c_j in which the current solution will remain optimal as well as feasible, the following condition should remain satisfied:

z coefficient of x_j = (LHS-RHS) of the jth dual constraint = (values of dual variables) * (coefficient vector a_j) – $c_j \geq 0$

Note: The previous condition is for a problem with maximization as an objective function. For the minimization objective, the condition will be ≤ 0. The z-row coefficient of a non-basic variable is also called its reduced cost, which represents the minimum amount by which an objective function will change (decrease) if one unit of that activity is to be produced.

To understand this in a better way consider the following example:

Example 6.2. *Primal Max z = 2x$_1$ + 2x$_2$ + x$_3$*

Subject to

$$x_1 + 5x_2 - x_3 \leq 50$$
$$2x_1 + x_2 + 3x_3 \leq 30$$
$$x_1, x_2, x_3 \geq 0$$

Dual   *Min w = 50y$_1$ + 30y$_2$*

Subject to

$$y_1 + 2y_2 \geq 2$$
$$5y_1 + y_2 \geq 2$$
$$-y_1 + 3y_2 \geq 1$$
$$y_1 \geq 0, y_2 \geq 0$$

Here y$_1$ and y$_2$ are the dual variables of the given problem.

Solution: It can be easily observed that s_1 = 50 and s_2 = 30 represent the initial basic feasible solution to the given problem. On solving the problem by using the simplex method, the following is the optimal table obtained:

Basic	x_1	x_2	x_3	s_1	s_2	Solution
x_2	0	1	– 0.556	0.222	– 0.111	7.778
x_1	1	0	1.778	– 0.111	0.556	11.111
z	0	0	1.444	0.222	0.889	37.778

Here x_1 and x_2 are the basic variables and x_3 is the non-basic variable. Also, we have y_1 = 0.222 and y_2 = 0.889.

Range of change of c_3: Since c_3 is the coefficient of the non-basic variable in the z equation, the allowable range in which c_3 can change without affecting the optimality of the final table is given by:

$$z\text{-row coefficient of } x_3 = -y_1 + 3y_2 - c_3^* = (0.222 \quad 0.889) * \begin{pmatrix} -1 \\ 3 \end{pmatrix} - c_3^* \geq 0$$

(the third constraint of the dual problem has been used)

$$\Rightarrow \quad -0.222 + 2.667 - c_3^* \geq 0$$

$$\Rightarrow \quad c_3^* \leq 2.446 \quad \Rightarrow \quad c_3 + \Delta c_3 \leq 2.446 \quad \Rightarrow \quad \Delta c_3 \leq 1.446$$

Here, c_3^* is the changed value of c_3.

This means the problem will remain optimal as well as feasible as long as the incremental value lies between $(-\infty, 1.446)$.

Here, if we change $c_3 = 1$ to $c_3 = 2 = c_3^*$ then we can see that

$$z \text{ row coefficient of } x_3 = (0.222 \quad 0.889) * \begin{pmatrix} -1 \\ 3 \end{pmatrix} - c_3^*$$

$$= -0.222 + 2.667 - 2 = 0.445$$

This is positive and still satisfying the optimality condition. Hence, there is no change in the current solution, because $\Delta c_3 = 1$, is inside the allowable range.

But if $c_3^* = 4 = c_3 + \Delta c_3 = 1 + 3$, that is, $\Delta c_3 = 3$, which is outside the allowable range, then we have

$$z \text{ row coefficient of } x_3 = (0.222 \quad 0.889) * \begin{pmatrix} -1 \\ 3 \end{pmatrix} - c_3^*$$

$$= -0.222 + 2.667 - 4 = -1.555$$

Now the optimality condition has been disturbed, which can be satisfied by applying the simplex method again, where x_3 will be the entering variable. Hence, we get the new solution as:

Basic	x_1	x_2	x_3	s_1	s_2	Solution
x_2	0.31271	1	0	0.18729	0.06287	11.2525
x_3	0.56243	0	1	-0.0624	0.31271	6.24916
z	0.87458	0	0	0.12492	1.37527	47.4954

Now the optimal solution is $x_2 = 11.2525$ and $x_3 = 6.24916$, $z = 47.4954$.

6.5.2. Changes in Objective Function Coefficients of Basic Variables

If the cost coefficients of some variables change in the objective function and these variables appear in the basis of the final simplex table, then the only changes in the final table will occur in the z-row coefficients of the non-basic variables. The z-row coefficients of the basic variable will still be zero, but these coefficients of the non-basic variables may or may not still satisfy the optimality condition. We need to recalculate them for non-basic variables as follows:

Changed (new) values of dual variables = (changed cost vector of basic variables in the final table) * (basis inverse in the final table)

z row coefficient of non-basic variable x_j = (LHS-RHS) of the jth dual constraint = *(New values of dual variables) * (original coefficient vector a)* $-c_j \geq 0$

*The new dual variables should be used in this case.

If these newly obtained z- row coefficients still satisfy the optimality condition, then the current solution will remain feasible as well as optimal; otherwise, the simplex method can be applied again to get the new optimal feasible solution.

Now again consider the Example (6.2).

Range of change of c_1: Let c_1^* be the new (changed) value of c_1, that is, $c_1^* = c_1 + \Delta c_1$.

The new values of dual variables = (changed cost vector of basic variables in the final table) * (basis inverse in the final table)

$$(y_1 \quad y_2) = \begin{pmatrix} 2 & c_1^* \end{pmatrix} \begin{pmatrix} 0.222 & -0.111 \\ -0.111 & 0.556 \end{pmatrix}$$

$$= \begin{pmatrix} 0.444 - 0.111c_1^* & -0.222 + 0.556c_1^* \end{pmatrix}$$

The new value of the z-row coefficient of the non-basic variable x_3

$$= \begin{pmatrix} 0.444 - 0.111c_1^* & -0.222 + 0.556c_1^* \end{pmatrix} * \begin{pmatrix} -1 \\ 3 \end{pmatrix} - c_3$$

For the current solution to remain optimal, we should have,

$$\Rightarrow \quad -0.444 + 0.111c_1^* + 3\left(-0.222 + 0.556c_1^*\right) - 1 \geq 0 \quad (\text{Since } c_3 = 1)$$

$$\Rightarrow \qquad\qquad\qquad\qquad\qquad c_1^* \geq 1.1861 \qquad\qquad ...(6.1)$$

The new value of the z row coefficient of the non-basic variable s_1

$$= \left(0.444 - 0.111 c_1^* - 0.222 + 0.556 c_1^*\right) * \binom{1}{0} - 0$$

For the current solution to remain optimal, we should have:

$$\Rightarrow \quad 0.444 - 0.111 c_1^* \geq 0$$
$$\Rightarrow \quad c_1^* \leq 4 \qquad \qquad \qquad \text{...(6.2)}$$

The new value of the z row coefficient of the non-basic variable s_2

$$= \left(0.444 - 0.111 c_1^* - 0.222 + 0.556 c_1^*\right) * \binom{1}{0} - 0$$

For the current solution to remain optimal, we should have:

$$\Rightarrow \quad -0.222 + 0.556 c_1^* \geq 0$$
$$\Rightarrow \quad c_1^* \geq 0.39928 \qquad \qquad \text{...(6.3)}$$

On combining (6.1), (6.2), and (6.3), the allowable range of c_1^* is given by

$$\text{Max}\{1.1861, 0.39928\} \leq c_1^* \leq 4$$

$$\Rightarrow \qquad 1.1861 \leq c_1^* \leq 4$$

As long as the changed value of c_1 lies within this range, the present solution will remain optimal.

Similarly, the allowable range of c_2 is $1 \leq c_2^* \leq 4.6$.

The following is the sensitivity report of the previous analysis obtained using WinQSB:

09-03-2016 09:30:30	Decision Variable	Solution Value	Reduced Cost	Unit Cost or Profit C(j)	Allowable Min. C(j)	Allowable Max. C(j)
1	X1	11.1111	0	2.0000	1.1875	4.0000
2	X2	7.7778	0	2.0000	1.0000	4.6000
3	X3	0	-1.4444	1.0000	-M	2.4444

Fig. 6.7

*M is a very large positive number.

Interpretations of the previous output:

- Allowable range for c_1 is between 1.1875 and 4, that is, $1.1875 \le c_1^* \le 4$.

- Allowable range for c_2 is between 1 and 4.6, that is, $1 \le c_2^* \le 4.6$.

- Allowable range for c_3 is between $-\infty$ and 2.444, that is, $-\infty \le c_3^* \le 2.4444$.

If we take $c_1^* = c_1 + \Delta c_1 = 3 = 2 + 1$ where $\Delta c_1 = 1$

then we have:

z row coefficient of the non-basic variable $x_3 = -2.11 + 1.779 c_1^* = 3.227$

z row coefficient of the non-basic variable $s_1 = -0.444 + 0.111 c_1^* = 0.111$

z row coefficient of the non-basic variable $s_2 = -0.222 + 0.556 c_1^* = 1.446$

Hence, optimality conditions are satisfied for all variables, and therefore there is no change in the optimal solution.

Now if we take $c_1^* = c_1 + \Delta c_1 = 6 = 2 + 4$ where $\Delta c_1 = 4$ then we will have following:

z row coefficient of the non-basic variable $x_3 = -2.11 + 1.779 c_1^* = 8.564$

z row coefficient of the non-basic variable $s_1 = -0.444 + 0.111 c_1^* = -0.222$

z row coefficient of the non-basic variable $s_2 = -0.222 + 0.556 c_1^* = 3.114$

Now the optimality condition of s_1 has been violated, and hence we will apply the simplex method to get a new solution, which will be satisfying the feasibility as well as the optimality condition. s_1 will be the new entering variable in the basis. From the optimal table of the original problem we have:

Basic Var.	x_1	x_2	x_3	s_1	s_2	Solution	Ratio
x_2	0	1	-0.556	0.222	-0.111	7.778	35.03
x_1	1	0	1.778	-0.111	0.556	11.111	
z	0	0	8.564	-0.222	3.114		,

s_1 enters and x_2 leaves.

Basic Var.	x_1	x_2	x_3	s_1	s_2	Solution
s_1	0	4.505	-2.505	1	-0.5	35.031

x_1	1	0.5	1.5	0	0.501	15
z	0	1	8.008	0	3.003	

Now the optimal solution is $x_1 = 15$, $x_1 = 1$, and the corresponding value of z is $6 * 15 + 2 * 0 = 90$.

6.5.3. 100% Rule for Making Simultaneous Changes in the Objective Function Coefficients

As we have applied this rule for simultaneous changes in b_i, the same can be applied here. We will calculate the change (either increase or decrease) in each coefficient as a percentage of allowable change (either increase or decrease) in that coefficient and then take the sum of all these percentages of change in all the coefficients. If the sum is less than 100%, then the current solution will remain feasible as well as optimal; otherwise, it may change.

In the Example (6.2), if we make changes in all the components simultaneously, that is, we take

$$c^* = (c_1^*, c_2^*, c_3^*)$$
$$= (c_1 + \Delta c_1, c_2 + \Delta c_2, c_3 + \Delta c_3)$$
$$= (2 + 0.5, 2 + 1, 1 + 0.5)$$
$$= (2.5, 3, 1.5)$$

now, if we calculate these percentages, we get:

$$\text{Allowable \% change in } c_1 = \frac{\text{Actual change in } c_1}{\text{Maximum Allowable change in } c_1} \times 100$$
$$= \frac{0.5}{2} \times 100 = 25\%$$

Similarly,

$$\text{Allowable \% change in } c_2 = \frac{\text{Actual change in } c_2}{\text{Maximum Allowable change in } c_2} \times 100$$
$$= \frac{1}{2.6} \times 100 = 38.46\%$$

$$\text{Allowable \% change in } c_3 = \frac{\text{Actual change in } c_3}{\text{Maximum Allowable change in } c_3} \times 100$$
$$= \frac{0.5}{1.4} \times 100 = 35.71\%$$

The sum of these percentages is 99.17, which is less than 100%, and hence the current solution will remain feasible as well as optimal. The following is the optimal table with the changed value of c.

Basic Var.	x_1	x_2	x_3	s_1	s_2	Solution
x_2	0	1	-0.556	0.222	-0.111	7.778
x_1	1	0	1.778	-0.111	0.556	11.111
z	0	0	1.278	0.389	1.056	51.111

Remark. Like in the case of b, a percentage sum more than 100% does not ensure that the optimality condition will be either disturbed or remain satisfied, whereas if it is less than 100%, then the current solution will remain feasible as well as optimal.

6.6. Changes in the a_{ij} Coefficients

6.6.1. Changes in a_{ij} Coefficients of Non-Basic Variables

If there are some changes in a_{ij} coefficients, for example, the a_j vector of the variable x_j, which is a non-basic variable in the final table, then the change in the final simplex table corresponding to this change will be in the vector corresponding to this variable and the z-row coefficient of this variable only, which can be recalculated as follows:

Changed Vector corresponding to non-basic variable x_j in the final table = (basis inverse in final table) * (changed vector a_j in the original problem)

z row coefficient of x_j = (values of dual variables) * (changed coefficient vector a_j in original problem) $- c_j$

OR z row coefficient of x_j = (value of C_B) * (changed vector corresponding to non-basic variable x_j in the final table) $-$ price c_j

If the optimality condition is still satisfied, then the current solution will remain feasible as well as optimal, otherwise by applying the simplex method again where the entering variable will be x_j, optimality can be regained.

For example, if we consider the Example (6.2) and replace $a_3 = \begin{pmatrix} -1 \\ 3 \end{pmatrix}$ by $a_3^* = \begin{pmatrix} 2 \\ 4 \end{pmatrix}$, then we have:

z row coefficient of x_3 = (values of dual variables) * (changed coefficient vector a_3 in original problem) $-c_3 = (0.222 \ 0.889) * \begin{pmatrix} 2 \\ 4 \end{pmatrix} - 1 = 3$.

OR The changed vector corresponding to non-basic variable x_3 in the final table

$$= \begin{pmatrix} 0.222 & -0.111 \\ -0.111 & 0.556 \end{pmatrix} \begin{pmatrix} 2 \\ 4 \end{pmatrix} = \begin{pmatrix} 0 \\ 2.002 \end{pmatrix}.$$

z row coefficient of x_3 = (value of C_B)* (changed vector corresponding to non-basic variable x_3 in the final table) – price $c_3 = (2 \ \ 2) \begin{pmatrix} 0 \\ 2.002 \end{pmatrix} - 1 = 3.002 \approx 3$.

Since the new value of $z = 3 > 0$, the optimality condition remains satisfied, and hence the original optimal solution will remain optimal for the changed problem.

Now if we change it to $a_3^* = \begin{pmatrix} -10 \\ 3 \end{pmatrix}$ then we have,

changed vector corresponding to non-basic variable x_3 in the final table

$$= \begin{pmatrix} 0.222 & -0.111 \\ -0.111 & 0.556 \end{pmatrix} \begin{pmatrix} -10 \\ 3 \end{pmatrix} = \begin{pmatrix} -2.553 \\ 2.778 \end{pmatrix}.$$

z row coefficient of x_3 $= (2 \ \ 2) * \begin{pmatrix} -2.553 \\ 2.778 \end{pmatrix} - 1 = -0.55$.

Now, since the optimality condition is disturbed, the final table of the changed problem will be calculated as follows:

Basic	x_1	x_2	x_3	s_1	s_2	Solution
x_2	0	1	-2.553	0.222	-0.111	7.778
x_1	1	0	2.778	-0.111	0.556	11.111
z	0	0	-0.55	0.222	0.889	

x_3 enters and x_1 leaves.

Basic	x_1	x_2	x_3	s_1	s_2	Solution
x_2	0.919	1	0	0.120	0.400	18.000

x_3	0.360	0	1	-0.040	0.200	4.000
z	0.199	0	0	0.200	0.100	

Since the optimality condition is now truly satisfied, the optimal solution of the changed problem is $x_2 = 18$ and $x_3 = 4$, and the corresponding value of z is 40.

The range of the a_{13} coefficient in vector a_3 can be calculated as follows:

Let $a_3 = \begin{pmatrix} -1 \\ 3 \end{pmatrix}$ be changed into $a_3^* = \begin{pmatrix} a_{13}^* \\ 3 \end{pmatrix}$

The changed vector corresponding to the non-basic variable x_3 in the final table = (basis inverse in final table) * (changed vector a_3 in the original problem).

$$= \begin{pmatrix} 0.222 & -0.111 \\ -0.111 & 0.556 \end{pmatrix} \begin{pmatrix} a_{13}^* \\ 3 \end{pmatrix} = \begin{pmatrix} 0.222a_{13}^* - 0.333 \\ -0.111a_{13}^* + 1.668 \end{pmatrix}$$

z row coefficient of x_3 = (values of c_B) * (changed coefficient vector a_3 in final table) $- c_3$

$$= (2 \quad 2) \begin{pmatrix} 0.222a_{13}^* - 0.333 \\ -0.111a_{13}^* + 1.668 \end{pmatrix} - 1 = 0.222 + 1.67a_{13}^*$$

For the solution to remain optimal, the allowable range for a_{13}^* is

$0.222 + 1.67\ a_{13}^* \geq 0 \implies a_{13}^* \geq -7.52253$

Similarly the range of the a_{23} coefficient in vector a_3 can be calculated as follows:

Let $a_3 = \begin{pmatrix} -1 \\ 3 \end{pmatrix}$ be changed into $a_3^* = \begin{pmatrix} -1 \\ a_{23}^* \end{pmatrix}$

The changed vector corresponding to the non-basic variable x_3 in the final table = (basis inverse in final table) * (changed vector a_3)

$$= \begin{pmatrix} 0.222 & -0.111 \\ -0.111 & 0.556 \end{pmatrix} \begin{pmatrix} -1 \\ a_{23}^* \end{pmatrix} = \begin{pmatrix} -0.222 - 0.111a_{23}^* \\ 0.111 + 0.556a_{23}^* \end{pmatrix}$$

z row coefficient of x_3 = (values of c_B) * (changed coefficient vector a_3 in final table) $-$ price c_3

$$= \begin{pmatrix} 2 & 2 \end{pmatrix} \begin{pmatrix} -0.222 - 0.111a_{23}{}^* \\ 0.111 + 0.556a_{23}{}^* \end{pmatrix} - 1 = -1.222 + 0.89a_{23}{}^*$$

For the solution to remain optimal, the allowable range for $a_{13}{}^*$ is

$$-1.222 + 0.89\, a_{23}{}^* \geq 0 \quad \Rightarrow \quad a_{23}{}^* \geq 1.373.$$

6.6.2. Changes in a_{ij} - Coefficients of Basic Variables

If changes occur in some or all a_{ij} coefficients of the vector a_j corresponding to the variable x_j which appear as a basic variable in the final simplex table, then we will recompute the following:

Changed column vector of basic variable x_j in the final table = (basis inverse in the final table) * (changed vector a_j in the original problem)

z row coefficient of x_j = (values of dual variables) * (changed coefficient vector a_j in original problem) $- c_j$

OR z row coefficient of x_j = (value of c_B) * (changed vector corresponding to non-basic variable x_j in the final table) – price c_j.

After making these changes in the final table, we have to check the required conditions which must be satisfied in any simplex table:

(i) The vectors corresponding to the basic variables should always be unit vectors, and they should form an identity matrix.

(ii) z-row coefficients of the basic variables should always be zero.

If these conditions are not satisfied, then first we apply the Gauss elimination method to satisfy them; otherwise, we check for the optimality condition. If the optimality condition has been disturbed, then apply the simplex method to get the new feasible optimal solution else current solution will remain optimal.

Changes in a_{ij} coefficients of basic variables: Let $a_1 = \begin{pmatrix} 1 \\ 2 \end{pmatrix}$ be changed into $a_1^* = \begin{pmatrix} a_{11}{}^* \\ 2 \end{pmatrix}$.

Changed column vector of basic variable x_1 in the final table = (basis inverse in the final table) * (changed vector a_1, that is, a_1^* in the original problem)

$$= \begin{pmatrix} 0.222 & -0.111 \\ -0.111 & 0.556 \end{pmatrix} \begin{pmatrix} a_{11}{}^* \\ 2 \end{pmatrix} = \begin{pmatrix} 0.222a_{11}{}^* & -0.222 \\ -0.111a_{11}{}^* & +1.112 \end{pmatrix}$$

z row coefficient of x_1 = (value of c_B) * (changed coefficient vector a_1 in the final table) – price c_1

$$= (2 \quad 2)\begin{pmatrix} 0.222a_{11} *-0.222 \\ -0.111a_{11} *+1.112 \end{pmatrix} - 2$$

$$= 0.222a_{11} *-0.222$$

Let $a_{11} * = 2$, then the changed column vector of the basic variable x_1 in the final table

$$= \begin{pmatrix} 0.222a_{11} *-0.222 \\ -0.111a_{11} *+1.112 \end{pmatrix} = \begin{pmatrix} 0.444-0.222 \\ -0.222+1.112 \end{pmatrix} = \begin{pmatrix} 0.222 \\ 089 \end{pmatrix}.$$

And the z row coefficient = 0.222 * 2 – 0.222 = 0.222.

Then the final table for Example 6.2 can be written as follows:

Basic	x_1	x_2	x_3	s_1	s_2	Solution
x_2	0.222	1	– 0.556	0.222	– 0.111	7.778
x_1	0.89	0	1.778	– 0.111	0.556	11.111
z	0.222	0	1.444	0.222	0.889	

We now convert the column corresponding to the basic variable x_1 into a unit vector by doing the normal pivoting so that the columns of x_1 and x_2 give us an identity matrix; the following is the resulting table:

Basic	x_1	x_2	x_3	s_1	s_2	Solution
x_2	0.0	1.0	–1.0	0.2	– 0.2	5.0
x_1	1.0	0.0	2.0.	– 0.1	0.6	12.5
z	0.0	0.0	1.0	0.2	0.8	

Since the optimality condition is now true, the optimal solution of the changed problem is x_1 = 12.5 and x_2 = 5, and the corresponding value of z is 35.

6.7. Deletion of a Variable

We may sometimes wish to eliminate/delete any variable from the given LPP after the solution has been obtained. There are two possible cases, depending on whether the variable belongs to the basis or not. We shall now consider both cases:

Case 1: If the variable to be removed is a non-basic in the final (optimal) table of the given problem, then the removal of this variable from the problem will not affect anything; that is, the optimal solution of the problem will not be changed by its removal. Hence, the current optimal solution will remain optimal for the changed problem also.

Case 2. If the variable is a part of the optimal basic feasible solution, then either the variable has a zero or non-zero value, in the optimal basis. If the variable has a zero value, its removal will have no impact on the final solution as in case 1. If the variable to be removed has positive value in the solution, then it will definitely have an impact on the final solution of the problem, and such a situation can be handled with a "sign-reversed dual simplex pivot" approach. This is called "sign-reversed" as the calculation starts off with a positive value of the variable, rather than a negative value which normally is the case in the dual simplex method. To delete this variable from the table, we first multiply the row of the deleting basic variable by -1 (except for the 1 for the basic variable, so that we still have an identity matrix from the present basic variables) in the optimal table of the given problem, and we can make it look like an ordinary dual simplex. Using the dual simplex algorithm, after the first pivot, we can remove the deleted variable from the problem. Continue the calculations as in the case of a normal dual simplex until the basic solution is feasible for the primal problem.

Example 6.3. *In Example 6.2, let the variable x_1 be removed from the given LPP and obtain the resulting optimum solution of the changed LPP.*

Solution: Multiplying by -1 in the x_1 row of the optimal table of Example 6.2, we have:

Basic	$-x_1$	x_2	x_3	s_1	s_2	Solution
x_2	0	1	-0.556	0.222	-0.111	7.778
$-x_1$	1	0	-1.778	0.111	-0.556	-11.111
z	0	0	1.444	0.222	0.889	37.778

Now we can apply the dual simplex algorithm, as follows:

Basic	$-x_1$	x_2	x_3	s_1	s_2	Solution
x_2	0	1	-0.556	0.222	-0.111	7.778
$-x_1$	1	0	-1.778	0.111	-0.556	-11.111
z	0	0	1.444	0.222	0.889	37.778
Ratio			-0.8125		-1.6	

x_1 leaves and x_3 enters

Basic	$-x_1$	x_2	x_3	s_1	s_2	Solution
x_2	-0.3125	1	0	0.1875	0.0625	11.25
x_3	-0.5625	0	1	-0.0625	0.3125	6.25
z	0.8125	0	0	0.3125	0.4375	28.75

Since all the entries in the z-row are greater than or equal to zero, the optimal solution is reached. Hence, the new optimal solution is $x_2 = 11.25$ and $x_3 = 6.25$, and the corresponding value of the objective function is 28.75.

6.8. Deletion of a Constraint

In some situations, we may like to delete a constraint of the given problem after the problems have been solved. In such a situation, we might have two different situations.

(i) If the slack variable corresponding to the constraint which is to be deleted has a positive value in the optimal solution, then its deletion leaves the optimal solution unchanged. This is because the constraint under consideration is ineffective in determining the optimal solution to the given problem.

(ii) If the slack variable corresponding to the constraint which is to be deleted has a zero value in the optimal solution, then the deletion of such a constraint will change the optimal solution of the problem. This type of constraint is said to be an active constraint. To obtain the new optimal solution in such a situation, we will first have to make this active constraint into an inactive one. To do this we will have to introduce the corresponding slack variable in the basis. Thereafter, we delete this inactive constraint.

Example 6.4. *To understand this in a better way, let us consider the LPP given in Example 2.8.*

$$\text{Maximize} = 4x_1 + x_2$$

Subject to the constraints:

$$2x_1 + x_2 \leq 100$$

$$x_1 + x_2 \leq 80$$

$$x_1 \leq 40$$

$$x_1, x_2 \geq 0$$

Solution: The following is the optimal table for this example.

Basic	x_1	x_2	s_1	s_2	s_3	Solution
x_2	0	1	1	0	−2	20
s_2	0	0	−1	1	1	20
x_1	1	0	0	0	1	40
z	0	0	1	0	2	180

Here $x_1 = 40$ and $x_2 = 20$ is the optimal solution of the given problem, and the corresponding optimal value z is 180.

Clearly, the second constraint is an inactive constraint, as its slack variable is at a positive level in the basis, and removing this constraint from the problem will not create any change in the present optimal solution. This can also be easily seen from the following graph:

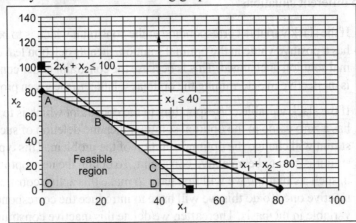

Fig. 6.8

In the previous graph, the corner point C (40, 20) gives us the optimal solution of the problem. Clearly removing $x_1 + x_2 \leq 80$ from the graph, is not going to change the optimal point. But on the other hand, removing either the first or third constraint from the problem will disturb the optimal point. The slack variables corresponding to the first and third constraints are at a zero level in the final solution (which can also be seen from the table).

Let us now remove the first constraint from the problem. As mentioned previously, to handle such situations, we will first have to make this constraint an inactive constraint by introducing the corresponding slack variable into the basis and thereafter deleting the first constraint, as follows:

Basic	x_1	x_2	s_1	s_2	s_3	Solution
x_2	0	1	1	0	−2	20
s_2	0	0	−1	1	1	20
x_1	1	0	0	0	1	40
z	0	0	1	0	2	180

Allow s_1 to enter the basis, and correspondingly x_2 leaves the basis.

Basic	x_1	x_2	s_1	s_2	s_3	Solution
s_2	0	1	1	0	−2	20
s_2	0	1	0	1	−1	40
x_1	1	0	0	0	1	40
z	0	−1	0	0	4	160

Now in the previous table s_1 is in the basis with a positive value 20, and hence the first constraint becomes inactive and can be removed from the problem. Thus, our problem becomes:

Maximize $z = 4x_1 + x_2$

Subject to the constraints:

$$x_1 + x_2 \leq 80$$
$$x_1 \leq 40$$
$$x_1, x_2 \geq 0$$

Now, the initial simplex table, to begin the iteration, is as follows:

Basic	x_1	x_2	s_1	s_2	s_3	Solution
s_2	0	1	0	1	−1	40
x_1	1	0	0	0	1	40
z	0	−1	0	0	4	160

Now we continue our simplex calculation to find the optimal solution.

Basic	x_1	x_2	s_1	s_2	s_3	Solution
x_2	0	1	0	1	−1	40
x_1	1	0	0	0	1	40
z	0	0	0	1	3	200

So, our new optimal solution is $x_1 = 40$ and $x_2 = 40$, and the correspond-ing maximum value of z is 200.

The following is the graphical solution to the changed problem:

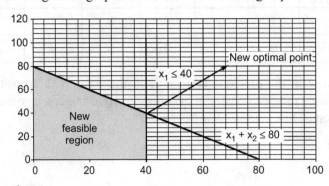

Fig. 6.9

6.9. Sensitivity by Using Excel Solver

Let us use Example 6.1.

(i) Enter the problem into an Excel sheet as follows:

		x1	x2					
objective coefficient		1	3					
		x1	x2	LHS	Relation	RHS		Slack
constraint co-efficients		1	0	0	<=	5		5
		1	2	0	<=	10		10
		0	1	0	<=	4		4
Solution		x1	x2	z				
				0				

(ii) Now enter all the necessary information of parameters in the Solver dialogue box as follows:

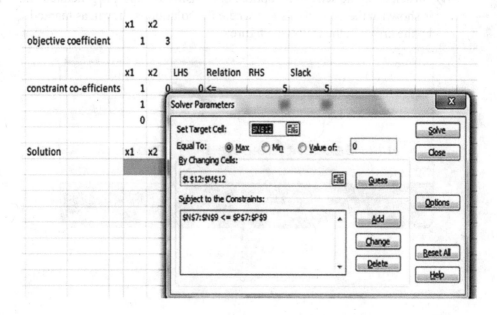

(iii) After entering all the information, click solve. We now will see the following window. Select the **sensitivity** option.

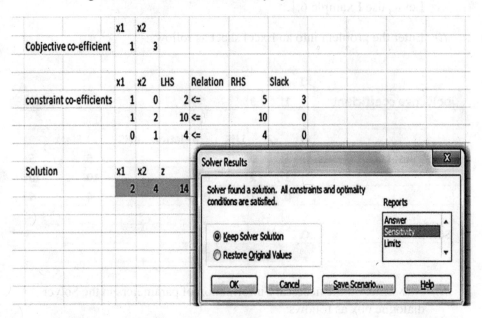

	x1	x2
Cobjective co-efficient	1	3

	x1	x2	LHS	Relation	RHS	Slack
constraint co-efficients	1	0	2	<=	5	3
	1	2	10	<=	10	0
	0	1	4	<=	4	0

Solution	x1	x2	z
	2	4	14

Solver Results

Solver found a solution. All constraints and optimality conditions are satisfied.

Reports

Answer
Sensitivity
Limits

⦿ Keep Solver Solution
○ Restore Original Values

OK   Cancel   Save Scenario...   Help

(iv) On selecting the sensitivity option, the sensitivity report is generated and is shown at the bottom of sheet where the solution is shown, as marked by the arrow in the following figure:

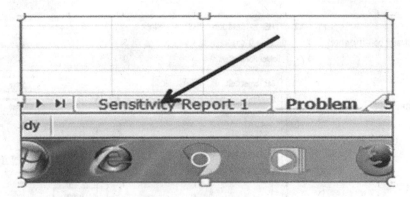

Sensitivity Report 1   Problem

dy

(v) On clicking the report button, we have the following sensitivity report:

Microsoft Excel 12.0 Sensitivity Report
Worksheet: [Book1]Sheet1
Report Created: 8/31/2016 8:57:20 AM
Adjustable Cells

Cell	Name	Final Value	Reduced Cost	Objective Coefficient	Allowable Increase	Allowable Decrease
H17 x1		2	0	1	0.5	1
I17 x2		4	0	3	1E+30	1

Constraints

Cell	Name	Final Value	Shadow Price	Constraint R.H. Side	Allowable Increase	Allowable Decrease
J12	Co-efficient in constraints LHS	2	0	5	1E+30	3
J13	LHS	10	1	10	3	1E+30
J14	LHS	4	1	4	1E+30	1.5

Solver sensitivity Report of Example 6.1.

Similarly, we also obtain the sensitivity report of Example 6.2 as follows:

Microsoft Excel 12.0 Sensitivity Report
Worksheet: [Book1]Sheet3
Report Created: 9/1/2016 6:04:25 PM

Adjustable Cells

Cell	Name	Final Value	Reduced Cost	Objective Coefficient	Allowable Increase	Allowable Decrease
H13		11.11111111	0	2	2	0.8125
I13		7.777777778	0	2	2.6	1
J13		0	-1.444444444	1	1.444444444	1E+30

Constraints

Cell	Name	Final Value	Shadow Price	Constraint R.H. Side	Allowable Increase	Allowable Decrease
K9		50	0.222222222	50	100	35
K10		30	0.888888889	30	70	20

Solver sensitivity Report of Example 6.2.

Exercises

1. What do you understand by the sensitivity analysis of a linear programming problem? Discuss in detail.

2. Discuss all the information given in the sensitivity report of a linear programming problem when it is solved by Excel Solver. Interpret it in detail.

3. What is the significance of the 100% rule in sensitivity analysis? Explain.

4. A baking company produces two types of cookies, Choco-chip cookies and Atta cookies, which require three types of processes for their production: mixing of ingredients, baking, and packaging. The time required by each unit of the two products and the maximum time available per week for the three processes are given as follows:

	Choco-chip Cookie(hr/ batch)	Atta Cookie (Hr/batch)	Max time available Per week(in hours)
Mixing	1.5	1	30
Baking	3	2	40
Packaging	2	2	35
Maximum Profit/batch	500	200	

Determine

(i) How many batches of both types of cookies should be produced per week so that the profit is at a maximum?

(ii) The range of variation of objective function coefficients in which the current optimal solution will not change.

(iii) The effect on the current optimal solution if the profit/batch on the two types of cookies gets changed to (a) $c = (350, 300)$ (b) $c = (400, 350)$. In both the cases, check the 100% rule also to verify the effect.

(iv) The range of variation of maximum time available/week for the three types of processes in which the current solution remains optimal. Explain why the allowable increases for mixing and packaging times are infinity.

(*v*)   The effect of deleting the third constraint from the problem. Also discuss the effect of deleting the second constraint from the problem.

(*vi*)   The effect of stopping the production of Atta cookies on the optimal solution and profit.

(*vii*)   The company is thinking of producing one more type of product, Besan-Rolls, for which the same three types of processing are required, and the processing time (in hr/batch) on the three operations are 2, 2.5, and 1 and profit/batch is $4.50/-. What will the change be in the optimal solution?

5.   Solve the following linear programming problem:

$$\text{Maximize } z = 2x_1 + 5x_2 + 8x_3$$

Subject to the constraints $2x_1 + 3x_2 + 4x_3 \leq 15$

$$2x_1 + 3x_2 + x_3 \leq 35$$

$$3x_1 + x_2 + 2x_3 \leq 40, \; x_1, x_2, x_3 \geq 0$$

Consider the following changes in the problem:

(*i*)   Determine the range of variation of different components of a_1 and a_2 so that the current solution remains unchanged.

(*ii*)   Determine the effect on the current optimal solution if $a_1 = \begin{pmatrix} 2 \\ 2 \\ 3 \end{pmatrix}$ changes to $a_1 = \begin{pmatrix} 3 \\ 2 \\ 4 \end{pmatrix}$ and $c_1 = 2$ to $c_1 = 7$.

(*iii*)   Determine the effect on the current optimal solution if $a_2 = \begin{pmatrix} 3 \\ 3 \\ 1 \end{pmatrix}$ changes to $a_1 = \begin{pmatrix} 3 \\ 5 \\ 3 \end{pmatrix}$

(*iv*)   Determine the effect on the current optimal solution if $a_3 = \begin{pmatrix} 4 \\ 1 \\ 2 \end{pmatrix}$ changes to $a_3 = \begin{pmatrix} 4 \\ 5 \\ 3 \end{pmatrix}$ and $c_3 = 8$ to $c_3 = 3$.

6. Solve the following linear programming problem:

$$\text{Maximize } z = 30x_1 + 45x_2$$

Subject to the constraints $5x_1 + 7x_2 \leq 60$

$$3x_1 + 2x_2 \leq 4, \; x_2, \; x_2 \geq 0$$

Determine the effect of the following changes on the current optimal solution:

(i) $c_1 = 30$ changes to $c_1 = 80$

(ii) $b = (60, 4)$ changes to $b = (15, 10)$

(iii) $x_1 \geq 1$ constraint is added to the problem

(iv) $x_2 \geq 1$ is added to the problem

CHAPTER 7

TRANSPORTATION AND TRANSSHIPMENT PROBLEMS

7.1. Introduction

The linear programming problems discussed in Chapters 2, 3, and 4 are all examples of classical "activity" models. In such models, the variables and constraints deal with distinctly different kinds of activities. This chapter introduces a significantly different but equally important model, in which something is shipped or assigned to the different locations. The resulting constraints represent both limitations on availability and requirements for delivery, and they have an especially simple form. One such problem is known as the transportation problem, in which a single good is to be shipped from several origins to several destinations at minimum overall cost. This problem gives rise to the simplest kind of linear program for minimum-cost flow problems.

Definition: A general transportation problem can be described as follows: a homogenous product or a commodity is available or stored in some quantities at a number of locations called origins (or sources) which have to be transported to various other locations called destinations. These destinations have their own demand which has to be fulfilled. Under the assumption that it is possible to transport from any origin to any other destination, the objective of the transportation problem is to determine the amount of the product which can be transported from each origin to the different destinations so that the total cost incurred in transportation to satisfy the demand at various destinations is minimized. In order to achieve this objective, it is assumed that (*i*) the amount and location of available supplies and the quantity demanded is known, and (*ii*) the unit cost of transportation from one origin to another destination is also known. A transportation problem is one of the applications of a linear programming problem in which requirements and resources are expressed in terms of only one kind of unit.

7.2. Formulation of a Transportation Problem

Let there be *m* origins and *n* destinations.

a_i – *Number of units of the commodity available at the source i, i = 1, 2 ... m*

b_j – *Number of units of the commodity required at the destination j, j = 1, 2 ... n*

c_{ij} – *Unit cost of transportation from origin i to the destination j*

x_{ij} – *Number of units or quantity transported from origin i to the destination j*

It is assumed that the total quantity available at the origins must be precisely the same as the total quantity demanded at the destinations, that is:

$$\sum_{i=1}^{m} a_i = \sum_{j=1}^{n} b_j$$

...(7.1)

The problem is to find $x_{ij} \geq 0$ ($i = 1, 2 ... m, j = 1, 2 ... n$) which satisfies the given ($m + n$) constraints and minimizes the total transportation cost, that is:

$$Min\ z = \sum_{i=1}^{m} \sum_{j=1}^{n} c_{ij} x_{ij}$$

...(7.2)

Subject to

$$\sum_{i=1}^{m} x_{ij} = b_j, j = 1, 2 \ ... n\,(Demand\ constraints)$$

...(7.3)

$$\sum_{j=1}^{n} x_{ij} = a_i, i = 1, 2 \ ... m\,(Supply\ constraints)$$

...(7.4)

$$x_{ij} \geq 0$$

...(7.5)

as it can be seen that it is formulated as a linear programming problem which has ($m + n - 1$) constraints (one constraint becomes redundant due to the assumption (7.1)) and ($m*n$) decision variables or unknowns.

The given transportation problem can be written in the matrix form like a standard LPP as follows:

$$Min\ z = (c_{11}x_{11} + ... + c_{1n}x_{1n}) + (c_{21}x_{21} + ... + c_{2n}x_{2n}) + ... + (c_{m1}x_{m1} + ... + c_{mn}x_{mn})$$

Subject to

$$x_{11} + x_{12} + ... + x_{1n} = a_1$$
$$x_{21} + x_{22} + ... + x_{2n} = a_2$$
$$\vdots \qquad \vdots \qquad \vdots$$
$$x_{m1} + x_{m2} + ... + x_{mn} = a_m$$
$$x_{11} + x_{21} + ... + x_{m1} = b_1$$
$$\vdots \qquad \vdots \qquad \vdots$$
$$x_{1n} + x_{2n} + ... + x_{mn} = b_n$$

Further, in matrix form it can be written as:

$$Min\ z = CX \text{ Subject to } AX = b, X \geq 0$$

where $C = (c_{11}, c_{12} ... c_{mn})$ is $(1 \times mn)$ row vector, $X = [x_{m1}, x_{m2}, ..., x_{mn}]$ is a $(mn \times 1)$ column vector, $b = [a_1, a_2, ..., a_m, b_1, b_2, ..., b_n]$ is a $((m + n) \times 1)$ column vector and

$$A = \begin{matrix} 1_n & \cdots & 0 \\ \vdots & \ddots & \vdots \\ 0 \quad 0 & \cdots & 1_n \\ I_n \quad I_n & & I_n \end{matrix} \quad \text{is a } ((m + n) \times mn) \text{ matrix.}$$

where 1_n is a nth order sum vector and I_n is a nth order identity matrix.

Any transportation problem which can be written as an LP problem can be solved by using the normal simplex algorithm, but due to the special structure of matrix **A,** it is not a very efficient method to solve it. However, transportation problems have a special mathematical structure that can be exploited to provide a streamlined approach to the general simplex calculation. This streamlined version of the simplex method is a systematic procedure for arriving at a solution to the given transportation problem. These approaches not only save computational time but also allow the solution to have an integer value if required, which otherwise is not possible using the normal simplex calculation.

Further, a transportation problem can be expressed in the tabular form as follows:

Table 7.1

		Destinations				
		D1	*D2*		*Dn*	*Supply*
Origins	O1	c_{11} x_{11}	c_{12} x_{12}	...	c_{1n} x_{1n}	a_1
	O2	c_{21} x_{21}	c_{22} x_{22}	...	c_{2n}	a_2
	⋮	⋮	⋮		⋮	⋮
	O*m*	c_{m1} x_{m1}	c_{m2} x_{m2}	...	c_{mn} x_{mn}	a_m
Demand	b_1	b_2		...	b_n	

Figure 7.1 gives a network representation of a typical transportation problem with *m* sources and *n* destinations.

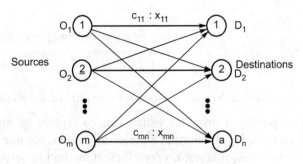

Fig. 7.1

Feasible Solution: A set of non-negative values of x_{ij}, $i = 1, 2 \ldots m$, $j = 1, 2 \ldots n$ which satisfies the constraints is called a feasible solution to the transportation problem.

Basic Feasible Solution: A feasible solution that contains no more than $(m + n - 1)$ non-negative allocations at independent positions is called a basic feasible solution to the transportation problem.

Non-Degenerate Basic Feasible Solution: A basic feasible solution that contains exactly $(m + n - 1)$ positive allocations at independent positions.

Degenerate Basic Feasible Solution: When the number of positive allocations in a given basic feasible solution are less than $(m + n - 1)$ then that solution is called a degenerate solution.

Optimal Solution: Any basic feasible solution which optimizes the objective function, that is, minimizes the total transportation cost is called an optimal solution.

Balanced and Unbalanced Transportation Problem: If the total supply in a transportation problem equals the total demand, then the transportation problem is called a balanced transportation problem; otherwise, it is an unbalanced transportation problem.

Theorem 7.1: The necessary and sufficient condition for the existence of the feasible solution to any transportation problem is:

$$\sum_{i=1}^{m} a_i = \sum_{j=1}^{n} b_j$$

Proof: (Necessary part) Let there exist a feasible solution $x_{ij} \geq 0$ to the transportation problem such that:

$$\sum_{i=1}^{m} x_{ij} = b_j \quad j = 1, 2, \ldots, n \quad \text{and} \quad \sum_{j=1}^{n} x_{ij} = a_i, \quad i = 1, 2, \ldots, m$$

$$\therefore \quad \sum_{j=1}^{n}\sum_{i=1}^{m} x_{ij} = \sum_{j=1}^{n} b_j \qquad \text{and} \qquad \sum_{i=1}^{m}\sum_{j=1}^{n} x_{ij} = \sum_{i=1}^{m} a_i$$

Hence $\qquad \sum_{i=1}^{m} a_i = \sum_{j=1}^{n} b_j$

(Sufficient Part) Let $\sum_{i=1}^{m} a_i = \sum_{j=1}^{n} b_j = M$ and $x_{ij} = \dfrac{a_i b_j}{M} \forall i, j$ ...(7.6)

Where M is positive constant. Hence, $x_{ij} \geq 0$ since $a_i > 0$, $b_j > 0 \ \forall \ i, j$

Now $\qquad \sum_{j=1}^{n} x_{ij} = \sum_{j=1}^{n} \dfrac{a_i b_j}{M} = a_i, i = 1, 2 \ldots m$ (from (7.6))

Similarly, $\qquad \sum_{i=1}^{m} x_{ij} = \sum_{i=1}^{m} \dfrac{a_i b_j}{M} = b_j, j = 1, 2 \ldots n$ (from (7.6))

Hence, x_{ij} satisfies all the constraints and therefore is a feasible solution for the transportation problem.

Theorem 7.2: The optimum solution of a transportation problem remains unchanged when a fixed constant is added or subtracted from every element of the given cost matrix.

Proof: Let,

$$\text{Min } z = \sum_{i=1}^{m} \sum_{j=1}^{n} c_{ij} x_{ij}$$

Subject to

$$\sum_{i=1}^{m} x_{ij} = b_j, j = 1, 2 \ldots n \,(\textit{Demand constraints}) \qquad \ldots(7.3)$$

$$\sum_{j=1}^{n} x_{ij} = a_i, i = 1, 2 \ldots m \,(\textit{Supply constraints}) \qquad \ldots(7.4)$$

$$x_{ij} \geq 0 \qquad \ldots(7.5)$$

be the given transportation problem. Let $c'_{ij} = c_{ij} \pm A$ where "A" is any positive constant and z' is the corresponding objective function. Then the previous problem can be rewritten as:

$$\text{Min } z' = \sum_{i=1}^{m} \sum_{j=1}^{n} c'_{ij} x_{ij}$$

$$= \sum_{i=1}^{m} \sum_{j=1}^{n} (c_{ij} \pm A) x_{ij}$$

$$= \sum_{i=1}^{m} \sum_{j=1}^{n} c_{ij} x_{ij} \pm A \sum_{i=1}^{m} \sum_{j=1}^{n} x_{ij}$$

$$= \sum_{i=1}^{m} \sum_{j=1}^{n} c_{ij} x_{ij} \pm A * M, \left(\text{where } M = \sum_{i=1}^{m} \sum_{j=1}^{n} x_{ij} \right)$$

$$= z \pm A * M$$

Subject to

$$\sum_{i=1}^{m} x_{ij} = b_j, j = 1, 2 \ldots n \,(\textit{Demand constraints}) \qquad \ldots(7.3)$$

$$\sum_{j=1}^{n} x_{ij} = a_i, i = 1, 2 \ldots m \,(\textit{Supply constraints}) \qquad \ldots(7.4)$$

$$x_{ij} \geq 0 \qquad \ldots(7.5)$$

We can clearly see that our new objective function is now equal to the original objective function $\pm A*M$ (constant term) (constant term) whereas the constraints remain the same. Thus, minimizing z' will give the same set of optimum solutions (allocations) as minimizing the original objective function.

7.3. Various Methods for Finding the Initial Basic Feasible Solution

The following are the methods for finding the initial basic feasible solution to a given transportation problem:

(i) North-West Corner method

(ii) Row minima method

(iii) Column minima method

(iv) Least-cost or matrix minima method

(v) Vogel's approximation method

The first method is "mechanical" in nature, and the remaining methods are heuristics which normally give a better starting solution compared to the North-West Corner rule. Of all the methods, Vogel's approximation method is considered to be the best one, as the initial basic feasible solution obtained by using this method is nearest to the optimum solution.

Note: Before we start finding initial basic feasible solutions, first make sure that the problem is a balanced one, that is, total supply is equal to total demand, which is a sufficient condition for finding the initial basic feasible solution.

7.3.1. North-West (N-W) Corner Method

In this method, the basic variables are selected from the North-West corner (*i.e.*, top left corner) of the tableau. We start by selecting the North-West corner cell (cell c_{11}) of the transportation table and allocate as many units as possible equal to the minimum between available supply and demand requirements, that is, min (a_1, b_1). Adjust the supply and demand numbers in the respective rows and columns accordingly. If the demand for the first column is fulfilled, then by crossing off the column, move horizontally to the next cell in the second column (North-West corner); otherwise, if the supply for the first row is exhausted, then by crossing off that row, move down to the first cell in the second row. If for any cell the demand equals the supply, all the units are allocated to the current cell, and by crossing off the corresponding row and column, move to the next North-West corner, which is the next row and next column, and make the allocation there. In this case when we cross off both row and column simultaneously, the total number of allocations will be less than $(m + n - 1)$ and we would have to generate more basic variables with zero allocations to make the total number of basic variables equal to $(m + n - 1)$. Hence, in order to avoid this situation, we can cross off either that row and move to the cell in the next row in the N - W corner by making demand in that column zero, or we can cross off the column and move to the cell in the next column in the N - W corner by making supply at that source zero. In both cases, however, the degenerate basic feasible solution will arise. This procedure will be continued until all the demands are fulfilled.

Note:

1. Total number of basic variables (or allocations) in the transportation problem must always be $(m + n - 1)$, where m and n represent the number of origins and destinations, respectively.

2. When the total supply is equal to the total demand at any cell allocation, then the degenerate basic feasible solution will arise (the condition of degeneracy).

Example 7.1. *Use the North-West-Corner rule to find an initial basic feasible solution to the following transportation problem.*

	D1	D2	D3	D4	D5	Supply(a)
O_1	2	2	3	4	2	20
O_2	5	6	3	7	2	35
O_3	3	2	9	4	7	40
Demand (b)	25	10	30	15	15	95

Solution: Here, $m = 3$, $n = 5$, and $\sum_{i=1}^{3} a_i = \sum_{j=1}^{5} b_j = 95$ (balanced T.P.), $(m + n - 1) = 7 = $ no. of basic variables where allocations will be $> = 0$, and others are non basic variables where allocation is zero.

	D_1	D_2	D_3	D_4	D_5	Supply
O_1	2 **20**	2	3	4	2	20
O_2	5 **5**	6 **→10**	3 **→20**	7	2	35
O_3	3	2	9 **10**	4 **→15**	7 **→15**	40
Demand	25	10	30	15	15	95

The initial basic feasible solution as per the previous calculation is $x_{11} = 20$, $x_{21} = 5$, $x_{22} = 10$, $x_{23} = 20$, $x_{33} = 10$, $x_{34} = 15$ and $x_{35} = 15$, and the corresponding cost is

$$z = 2 \times 20 + 5 \times 5 + 6 \times 10 + 3 \times 20 + 9 \times 10 + 4 \times 15 + 7 \times 15 = 440$$

7.3.2. Row Minima Method

In this method, the allocation starts with the first row. The minimum cost in the first row is identified (ties are broken arbitrarily) and the maximum possible units are shipped, which is the minimum of the first-row availability and corresponding column demand, that is, min (a_1, b_j), where b_j is the demand of the corresponding column. If the minimum occurs corresponding to the column demand b_j, then the column is crossed off and the next minimum of the same row is identified and the same procedure is repeated. If the minimum occurs corresponding to the row, that is, $a_1 < b_j$, then we move to the next row by crossing off the current row and follow the same procedure again until all the requirements are satisfied.

Example 7.2. *Use the row minima method to find an initial basic feasible solution to the transportation problem given in Example 7.1.*

Solution: The minimum in the first row occurs at three different locations, namely, $(1, 1)$, $(1, 2)$ and $(1, 5)$. Arbitrarily selecting $(1, 1)$, we allocate the maximum possible number of units in this cell, which is min $(20, 25) = 20$ and is corresponding to the availability in the row. Now, the first row supply is exhausted, and the column demand is accordingly adjusted. Now we move to the second row.

	D1	D2	D3	D4	D5	Supply
O₁	2 **20**	2	3	4	2	~~20~~
O₂	5	6	3	7	2	35
O₃	3	2	9	4	7	40
Demand	~~25~~ 5	10	30	15	15	95

Allocate 15 at (2, 5). Demand for column 5 is fulfilled and supply adjusted accordingly. We will still remain in the second row.

	D1	D2	D3	D4	D5	Supply
O₁	2 **20**	2	3	4	2	~~20~~
O₂	5	6	3	7	2 **15**	~~35~~ 20
O₃	3	2	9	4	7	40
Demand	~~25~~ 5	10	30	15	~~15~~	95

The next minimum in the second row is 3, which occurs at (2, 3); allocate 20 in this cell. The second row supply is exhausted and demand for the third column is adjusted accordingly. Move to the third row.

	D1	D2	D3	D4	D5	Supply
O₁	2 **20**	2	3	4	2	~~20~~
O₂	5	6	3 **20**	7	2 **15**	~~35 20~~
O₃	3	2	9	4	7	40
Demand	~~25~~ 5	10	~~30~~ 10	15	~~15~~	95

The minimum in the third row is at (3, 2); we allocate 10 in this. Demand for the second column is done, and the supply is adjusted accordingly.

	D1	D2	D3	D4	D5	Supply
O₁	2 20	2	3	4	2	~~20~~
O₂	5	6	3 20	7	2 15	~~35~~ 20
O₃	3	2 10	9	4	7	~~40~~ 30
Demand	~~25~~ 5	10	~~30~~ 10	15	~~15~~	95

The next minimum in the third row is with (3, 1); allocate 5 in this cell. Demand for column 1 is done now, and accordingly the supply is adjusted.

	D1	D2	D3	D4	D5	Supply
O₁	2 20	2	3	4	2	~~20~~
O₂	5	6	3 20	7	2 15	~~35~~ 20
O₃	3 5	2 10	9	4	7	~~40~~ ~~30~~ 25
Demand	~~25~~ 5	~~10~~	~~30~~ 10	15	~~15~~	95

The next minimum in the third row is with (3,4); allocate 15 in this cell. Demand for column 4 is done now, and accordingly the supply is adjusted.

	D1	D2	D3	D4	D5	Supply
O₁	2 20	2	3	4	2	~~20~~
O₂	5	6	3 20	7	2 15	~~35~~ 20
O₃	3 5	2 10	9	4 15	7	~~40~~ ~~30~~ ~~25~~ 10
Demand	~~25~~ 5	~~10~~	~~30~~ 10	~~15~~	~~15~~	95

The only cell left in the third row is (3, 3); allocate 10 in this. Demand for column 3 is also satisfied. It can be easily seen that all the demands at the different locations are now fulfilled, and we have the initial basic feasible solution to the given problem using this method.

	D1	D2	D3	D4	D5	Supply
O_1	2 20	2	3	4	2	~~20~~
O_2	5	6	3 20	7	2 15	~~35~~ 20
O_3	3 5	2 10	9 10	4 15	7	~~40~~ ~~30~~ 25 ~~10~~
Demand	~~25~~ 5	~~10~~	~~30~~ ~~10~~	~~15~~	~~15~~	95

The initial basic feasible solution as per the previous calculation is

$x_{11} = 20, x_{23} = 20, x_{25} = 15, x_{31} = 5, x_{32} = 10, x_{33} = 10$ and $x_{34} = 15$

The total cost of transportation using this method is

$z = 2 \times 20 + 3 \times 2 + 2 \times 15 + 3 \times 5 + 2 \times 10 + 9 \times 10 + 4 \times 15 = 315$

7.3.3 Column Minima Method

This is just the opposite of the row minima method. Instead of moving from one row to another, here starting with the first column, we move from one column to another column till until all the demands are satisfied.

Example 7.3. *Use the column minima method to find an initial basic feasible solution to Example 7.1.*

Solution: Following the same procedure as in the row minima method, we will move column-wise. We have the following table representing the initial basic feasible solution obtained using this method.

	D_1	D_2	D_3	D_4	D_5	Supply
O_1	2 20	2	3	4	2	20
O_2	5	6	3 30	7	2 5	35

	3	2	9	4	7	
O_3	5	10		15	10	40
Demand	25	10	30	15	15	95

We have

$$x_{11} = 20, x_{23} = 30, x_{25} = 5, x_{31} = 5, x_{32} = 10, x_{34} = 15 \text{ and } x_{35} = 10$$

and the total cost of transportation using this method is

$$z = 2 \times 20 + 3 \times 5 + 2 \times 10 + 3 \times 30 + 4 \times 15 + 2 \times 5 + 7 \times 10 = 305$$

7.3.4 Least-Cost or Matrix Minima Method

This method usually provides a better initial basic feasible solution than the North-West Corner method, and it's a generalization of the previous two methods, since it takes into account the minimum cost present in the entire tableau instead of looking for minimum cost in a single row or a single column at a time, by moving from one row to another row or from one column to another column, respectively. The method starts by assigning the maximum possible amount to the cell with the smallest unit cost in the entire tableau (ties are broken arbitrarily). Cross out either the row or column which is exhausted completely. If both (row as well as the column) are exhausted completely then also cross out only one of them by making the entry at either supply or demand equal to zero (to avoid the situation of having the no. of allocations to be less than $(m + n - 1)$ as discussed earlier). An adjustment of the supply and demand for the row and column which has not been crossed out is to be done. Next, select the uncrossed-out cell with the smallest unit cost and repeat the process until no row or column is left uncrossed out.

Example 7.4. *Use the least-cost method to find an initial basic feasible solution to Example 7.1.*

Solution: The least cost in the table is 2, which occurs at five different places; arbitrarily selecting the cell (1, 1), we put the maximum allocation possible in this.

	D1	D2	D3	D4	D5	Supply
O_1	2 20	2	3	4	2	20

O$_2$	5	6	3	7	2	35
O$_3$	3	2	9	4	7	40
Demand	~~25~~ 5	10	30	15	15	95

Now the first row supply is exhausted, and hence it has been stricken out and the demand for the first column is accordingly adjusted. Find the next minimum in the reduced table, which again is 2. The minima occurs at two different locations; arbitrarily selecting the cell (2, 5), the next allocation is done and the reduced table is shown as follows:

	D1	D2	D3	D4	D5	Supply
O$_1$	2 20	2	3	4	2	~~20~~
O$_2$	5	6	3	7	2 15	~~35~~ 20
O$_3$	3	2	9	4	7	40
Demand	~~25~~ 5	10	30	15	~~15~~	95

The demand for column 5 is fulfilled, and thus it's crossed out. Now the next minimum is at (3, 2). Put the maximum possible allocation in this cell. The reduced table is shown as follows:

	D1	D2	D3	D4	D5	Supply
O$_1$	2 20	2	3	4	2	~~20~~
O$_2$	5	6	3	7	2 15	~~35~~ 20
O$_3$	3	2 10	9	4	7	~~40~~ 30
Demand	~~25~~ 5	~~10~~	30	15	~~15~~	95

Demand for column 2 is satisfied now. The next minimum is 3, which is occurring at two different locations. Arbitrarily selecting the cell (2,3) and allocating the maximum possible units in this cell, we get:

	D1	D2	D3	D4	D5	Supply
O₁	2 20	2	3	4	2	~~20~~
O₂	5	6	3 20	7	2 15	~~35~~ ~~20~~
O₃	3	2 10	9	4	7	~~40~~ 30
Demand	~~25~~ 5	~~10~~	~~30~~ 10	15	~~15~~	95

The next allocation will be at cell (3, 1), and we have:

	D1	D2	D3	D4	D5	Supply
O₁	2 20	2	3	4	2	~~20~~
O₂	5	6	3 20	7	2 15	~~35~~ ~~20~~
O₃	3 5	2 10	9	4	7	~~40~~ ~~30~~ 25
Demand	~~25~~ 5	~~10~~	~~30~~ 10	~~15~~	~~15~~	95

The minimum cost cell is (3, 4) in the remaining uncrossed out table, so we have:

	D1	D2	D3	D4	D5	Supply
O₁	2 20	2	3	4	2	~~20~~
O₂	5	6	3 20	7	2 15	~~35~~ ~~20~~
O₃	3 5	2 10	9	4 15	7	~~40~~ ~~30~~ ~~25~~ 10
Demand	~~25~~ 5	~~10~~	~~30~~10	~~15~~	~~15~~	95

At last doing allocation at cell (3, 3), we get:

	D1	D2	D3	D4	D5	Supply
O₁	2 20	2	3	4	2	~~20~~
O₂	5	6	3 20	7	2 15	~~35~~ ~~20~~
O₃	3 5	2 10	9 10	4 15	7	~~40~~ ~~30~~ ~~25~~ ~~10~~
Demand	~~25~~ 5	~~10~~	~~30~~ ~~10~~	~~15~~	~~15~~	95

The initial basic feasible solution using the least-cost method as per the previous calculation is $x_{11} = 20, x_{23} = 20, x_{25} = 15, x_{32} = 10, x_{33} = 10$, and $x_{34} = 15$. The total cost of transportation using this method is $z = 2 \times 20 + 3 \times 20 + 2 \times 15 + 3 \times 5 + 2 \times 10 + 14 \times 15 = 315$

7.3.5 Vogel's Approximation Method (VAM)

This method not only takes into account the least cost but also the next minimum cost for each row and each column. The VAM method is an improved version of the least-cost method that generally, but not always, produces the best starting solution. The following steps are involved in this method:

Step 1: Determine the difference (penalty) between the minimum and the next minimum cost for each row and column (including dummies) and place it alongside in a separate column and row of the transportation table.

Step 2: Identify the row or column with the largest difference (penalty).

Step 3: Allocate as much as possible to the lowest-cost cell of the corresponding row or column with the highest penalty (in case of tie select the one which has least cost cell) and cross out either the row or column where supply or demand is exhausted completely. If it occurs for both, then also only one will be crossed out and the entry at the other will be made to be zero.

Step 4: Stop the process if all column requirements are met. If not, then go to the next step.

Step 5: Recalculate the difference between minimum and next minimum cost out of the remaining uncrossed out cells in all rows and columns. Any row and column with an exhausted supply or fulfilled demand should not be used in calculating further differences. Then go to Step 2.

Step 6. If only one row or column is left, with a positive supply, then apply the least-cost method to that row or column and complete the allocation.

Vogel's approximation method (VAM) usually produces an optimal or near optimal starting solution. Of all the five methods, this is the most widely accepted method for finding an initial basic feasible solution to a given transportation problem.

Example 7.5. *Use VAM to find an initial basic feasible solution to Example 7.1.*

Solution: Determine the difference (penalty) between the minimum and the next minimum cost for each row and each column (including dummies) and place it alongside the transportation table, as shown in the following table:

	D1	D2	D3	D4	D5	Supply	Penalty
O_1	2	2	3	4	2	20	0
O_2	5	6	3	7	2 **15**	~~35~~ 20	1
O_3	3	2	9	4	7	40	1
Demand	25	10	30	15	~~15~~		
Penalty	1	0	0	0	0		

The highest penalty occurs at three different locations (2nd and 3rd row and 1st column); arbitrarily selecting the second row, the minimum c_{ij} in this row is c_{25} (*i.e.*, 2). So, $x_{25} = 15$ and the fifth column is eliminated.

Again penalties are calculated for the new table.

	D1	D2	D3	D4	Supply	Penalty
O_1	2	2	3	4	20	0
O_2	5	6	3 **20**	7	20	2
O_3	3	2	9	4	40	1

Demand	25	10	~~30~~ 10	15		
Penalty	1	0	0	0		

The highest penalty occurs in the second row, where the minimum c_{ij} is c_{23} (*i.e.*, 3). So, $x_{23} = 20$ and the supply for the second row is exhausted, and hence it's eliminated.

New penalties are again calculated for the new reduced table as follows:

	D1	D2	D3	D4	Supply	Penalty
O_1	2	2	3 **10**	4	~~20~~ 10	0
O_3	3	2	9	4	40	1

Demand	25	10	~~30~~ ~~10~~	15
Penalty	1	0	6	0

The highest penalty occurs in the third column, where the minimum c_{ij} is c_{13} (*i.e.*, 3). So, $x_{13} = 10$, the demand for the third column is satisfied completely, and hence it is eliminated.

Now the reduced table has two supply and three demand points, and new penalties are calculated again as shown in the following table:

	D1	D2	D4	Supply	Penalty
O_1	2	2	4	10	0
O_3	3	2 **10**	4	~~40~~ 30	1
Demand	25	~~10~~	15		
Penalty	1	0	0		

The highest penalty occurs at two different locations, namely, the second row and first column; arbitrarily selecting the second row, the minimum c_{ij}

is c_{32} (*i.e.*, 2). So, $x_{32} = 10$. The second column of the reduced table is now deleted.

The highest penalty for the new table is 2, and this occurs in the first row of the reduced table; allocating the maximum possible number to the minimum element of this row, we have $x_{11} = 10$. In doing so the supply of the first row is now exhausted, and hence it's now eliminated.

	D1	D4	Supply	Penalty
O_1	2 10	4	~~10~~	2
O_3	3	4	30	1
Demand	~~25~~ 15	15		
Penalty	1	0		

Finally, we are left with only one row, and we apply the least-cost method to have the final allocations, as shown in the following:

	D1	D4	Supply
O_3	3 15	4 15	~~30~~ 15
Demand	~~15~~	~~15~~	

Combining all the previous information, the following gives the initial basic feasible solution to the given problem using Vogel's approximation method:

	D1	D2	D3	D4	D5	Supply
O_1	2 10	2	3 10	4	2	20
O_2	5	6	3 20	7	2 15	35
O_3	3 15	2 10	9	4 15	7	40
Demand	25	10	30	15	15	

The initial basic feasible solution using VAM as per the previous calculation is:

$x_{11} = 10$, $x_{13} = 10$, $x_{13} = 20$, $x_{25} = 15$, $x_{31} = 15$, $x_{32} = 10$, and $x_{34} = 15$

The total cost of transportation using this method is:

$z = 2 \times 10 + 3 \times 10 + 3 \times 20 + 2 \times 15 + 3 \times 15 + 2 \times 10 + 4 \times 15 = 265$

7.4. Closed Path or Loop in Transportation Problems

A closed path or a loop is a sequence of cells in the transportation table where, starting with a non-basic cell/empty cell (where there is no allocation), we move to the other basic/occupied cells and then come back to the starting cell with the following conditions:

(i) Every row or column should have exactly one pair of occupied cells at the corners of the loop.

(ii) The first and last cells of a loop lie in the same row or column.

(iii) No duplication of a cell is allowed; that is, no cell appears more than once in a loop.

(iv) Only horizontal and vertical moves are allowed and can only change directions at occupied cells.

(v) This concept of the closed path will be required in the optimality test while finding the solution to the transportation problem.

Note: All the allocations in the transportation problem occur at independent locations (i.e., using only the allocated cells, one will not be able to form a closed loop).

7.5. Moving Toward the Optimal Solution

In the previous section, we have seen the different methods for finding an initial basic feasible solution to a given transportation problem. After having constructed an initial basic feasible solution, our next step is to move toward an optimal solution. There are two methods for doing this:

(i) Stepping-stone method

(ii) Modified distribution (MODI) or u-v method

7.5.1. Stepping-Stone Method

This method has been derived from the analogy of crossing a pond using stepping stones. This means that the entire transportation table is assumed to be a pond, and the occupied cells (where allocations have been made) are the stones which are needed to make certain movements within the pond. The following steps are involved in this method:

(i) Find the initial basic feasible solution to the given transportation problem using any of the five methods discussed previously.

(ii) Starting with any non-basic cell, construct a loop via basic cells (or occupied cells).

(iii) For the given loop, begin with a plus (+) sign at the starting non-basic cell and alternately put minus (−) and plus (+) signs at each basic cell appearing at the corner point of the loop.

(iv) Now we calculate an Improvement Index by first adding the unit-cost figures found in each cell containing a plus sign and subtracting the unit costs in each cell containing a minus sign.

(v) Repeat steps *(ii)*-*(iv)* for each non-basic cell.

(vi) If the improvement indices calculated for all non-basic cells are greater than or equal to zero, then the present basic feasible solution is the optimal basic feasible solution. But if there is at least one index with a negative value, select the loop that has the most negative value (ties are broken arbitrarily) and further improves the solution.

(vii) To improve the current solution further, select the "smallest" value of the basic variables appearing at the corners of the loop and containing minus (−) signs. This number is added to all the cells appearing at the corner of the closed loop with a plus (+) signs and subtracted from all cells on the path assigned with a minus (−) sign, and we get an improved basic feasible solution.

(viii) Repeat steps *(i)* through *(vi)* until all improvement indices computed are greater than or equal to zero.

Example 7.6. *Apply the stepping-stone method to the initial basic feasible solution (obtained using VAM) of the transportation problem given in Example (7.1).*

	D1	D2	D3	D4	D5	Supply
O_1	2 10	2	3 10	4	2	20
O_2	5	6	3 20	7	2 15	35
O_3	3 15	2 10	9	4 15	7	40
Demand	25	10	30	15	15	95

The previous table has a total of 8 unoccupied cells, and hence we need to compute a total of eight improvement indexes by constructing a loop for each of these cells.

For the cell (1, 2):

	D_1	D_2	D_3	D_4	D_5	Supply
O_1	2 $-$ ↑10	2 +	3 10	4	2	20
O_2	5	6	3 20	7	2 15	35
O_3	3 + 15	2 $-$ ↓10	9	4 15	7	40
Demand	25	10	30	15	15	95

At the cell (1, 4):

	D_1	D_2	D_3	D_4	D_5	Supply
O_1	2 $-$ ↑10	2	3 10	4 +	2	20
O_2	5	6	3 20	7	2 15	35
O_3	3 15 +	2 10	9	4 15 ↓ $-$	7	40
Demand	25	10	30	15	15	95

At the cell (1, 5):

	D_1	D_2	D_3	D_4	D_5	Supply
O_1	2 10	2	− 3 ↑10	4	+ 2 	20
O_2	5	6	+ 3 20	7	− 2 15↓	35
O_3	3 15	2 10	9	4 15	7	40
Demand	25	10	30	15	15	95

At the cell (2, 1):

	D_1	D_2	D_3	D_4	D_5	Supply
O_1	− 2 ↑10	2	+ 3 10→	4	2	20
O_2	+ 5 	6	− 3 20	7	2 15	35
O_3	3 15	2 10	9	4 15	7	40
Demand	25	10	30	15	15	95

At the cell (2, 2):

	D_1	D_2	D_3	D_4	D_5	Supply
O_1	− 2 10	2	+ 3 →10	3	2	20
O_2	5	+ 6	− 3 20	7	2 15	35
O_3	+ 3 15	− 2 10	9	4 15	7	40
Demand	25	10	30	15	15	95

At the cell (2, 4):

	D_1	D_2	D_3	D_4	D_5	Supply
O_1	− 2 10	2	+ 3 10	3	2	20
O_2	5	6	− 3 20	+ 7	2 15	35
O_3	+ 3 15	2 10	9	− 4 15	7	40
Demand	25	10	30	15	15	95

At the cell (3, 3):

	D_1	D_2	D_3	D_4	D_5	Supply
O_1	+ 2 10	2	− 3 10	4	2	20
O_2	5	6	3 20	7	2 15	35
O_3	− 3 15	2 10	+ 9	4 15	7	40
Demand	25	10	30	15	15	95

Finally, at the cell (3, 5):

	D_1	D_2	D_3	D_4	D_5	Supply
O_1	+ 2 10	2	− 3 10	4	2	20
O_2	5	6	+ 3 20	7	− 2 15	35
O_3	− 3 15	2 10	9	4 15	+ 7	40
Demand	25	10	30	15	15	95

Now the improvement indexes at these locations are:

(i) $At (1, 2) = 2 + 3 - 2 - 2 = 1$

(ii) $At (1, 4) = 4 + 3 - 2 - 4 = 1$

(iii) $At (1, 5) = 2 + 3 - 3 - 2 = 0$

(iv) $At (2, 1) = 5 + 3 - 2 - 3 = 3$

(v) $At (2, 2) = 3 + 6 + 3 - 2 - 3 - 4 = 3$

(vi) $At (3, 3) = 9 + 2 - 3 - 3 = 5$

(vii) $At (3, 5) = 7 + 2 - 3 - 3 - 3 - 2 = 4$

Since all the improvement indexes are ≥ 0, the present basic feasible solution is the optimal basic feasible solution.

Hence, $x_{11} = 10$, $x_{23} = 10$, $x_{23} = 20$, $x_{25} = 15$, $x_{31} = 15$, $x_{32} = 10$, and $x_{34} = 15$ is the optimal basic feasible solution, and the corresponding total minimum cost of transportation is:

$$z = 2 \times 10 + 3 \times 10 + 3 \times 20 + 2 \times 15 + 3 \times 15 + 2 \times 10 + 4 \times 15 = 265$$

Example 7.7. *The following table gives the initial basic feasible solution for a transportation problem of a company, obtained using the North-West corner rule. Find the optimum transport cost for the company using the stepping-stone Method.*

market

		P	Q	R	Supply
	A	6 16	3 6	5	22
Warehouse	B	5	9 6	2 9	15
	C	5	7	8 8	8
	Demand	16	12	17	45

Solution: The total cost of transportation for the given initial basic feasible solution is $z = 250$. In the previous table, we have a total of five unoccupied cells. We will now form the closed loop and compute the improvement index for all these cells.

At the cell (1, 3):

market

		P	Q	R	Supply
		6	− 3	+ 5	
	A	**16**	**6**		22
Warehouse	B	5	− 9	− 2	15
			6	**9**	
	C	5	7	8	8
				8	
	Demand	16	12	17	45

At the cell (2, 1):

market

		P	Q	R	Supply
		− 6	+ 3	5	
	A	**16**	**6**		22
		+ 5	− 9	2	
Warehouse	B		**6**	**9**	15
	C	5	7	8	8
				8	
	Demand	16	12	17	45

At the cell (3, 1):

market

		P	Q	R	Supply
		− 6	+ 3	5	
	A	**16**	**6**		22
		5	− 9	+ 2	
Warehouse	B		**6**	**9**	15
	C	+ 5	7	− 8	8
				8	
	Demand	16	12	17	45

At the cell (3, 2):

market

	P	Q	R	Supply
A	6 **16**	3 6	5	22
Warehouse B	5	− 9 ▲**6**	+ 2 **9**	15
C	5	+ 7 **◄**	− ▼ 8 **8**	8
Demand	16	12	17	45

The improvement indices at these cells are:

(i) $At (1, 3) = 5 + 9 - 2 - 3 = 9$

(ii) $At (2, 1) = 5 + 3 - 6 - 9 = -7$

(iii) $At (3, 1) = 5 + 3 + 2 - 6 - 9 - 8 = -13$

(iv) $At (3, 2) = 7 + 2 - 9 - 8 = -8$

Since all the improvement indices are not ≥ 0, the current basic feasible solution is not an optimal basic feasible solution. As the most negative improvement index occurs with the (3, 1) cell, we allow this not-basic cell to enter the basis. We select the loop corresponding to the non-basic cell (3, 1) and use it to further improve the solution. In this loop, select the "smallest" value of the basic variables containing a minus (−) sign, which is 6 here. This number is added to all the cells on the closed loop with a plus (+) sign and subtracted from all cells assigned with a minus (−) sign. Thus, we have an improved basic feasible solution.

market

	P	Q	R	Supply
A	6 **10**	3 **12**	5	22
Warehouse B	5	9	2 **15**	15
C	5 **6**	7	8 **2**	8
Demand	16	12	17	45

Now the new improved basic variables are $x_{11} = 10, x_{12} = 12, x_{23} = 15, x_{31} = 6$, and $x_{33} = 2$, and the corresponding total cost of transportation is $z = 172$.

The previous steps are repeated again and again until we have the following optimal transportation solution:

market

		P	Q	R	Supply
Warehouse	A	6 **8**	3 **12**	5 **2**	22
	B	5	9	2 **15**	15
	C	5 **8**	7	8	8
	Demand	16	12	17	45

Here, $x_{11} = 8$, $x_{12} = 12$, $x_{13} = 2$, $x_{23} = 15$, and $x_{31} = 8$ is the optimal basic feasible solution, and the corresponding total minimum cost of transportation is $z = 164$.

7.5.2 The Modified (MODI) Distribution or u-v Method

We have seen that in every iteration of the stepping-stone method, one has to construct a stepping-stone path (or loop) for every non-basic cell. This aspect of the method is not only time-consuming but also laborious. The amount of effort involved also grows exponentially with the size of the problem. This is a very serious drawback of this method. It is, therefore, desirable to look for a more efficient alternative. The Modified (MODI) distribution or u-v method is one alternative whose computational effort grows only linearly. This method allows us to compute improvement indices quickly for each non-basic cell without drawing all of the closed paths. Because of this, the Modified (MODI) distribution method provides considerable time saving over the stepping-stone method. The MODI method provides a new means of finding the improvement index for the non-basic cells. Once the cell is identified, we are required to trace only one closed path.

To understand this method in a better way, we write the dual of the transportation problem given in (7.2). Rewriting (7.2), we have:

$$Min\, Z = \sum_{i=1}^{m} \sum_{j=1}^{n} c_{ij} x_{ij}$$

Subject to

$$\sum_{i=1}^{m} x_{ij} \ge b_j, \qquad j = 1, 2, \dots, n$$
$$\left. \sum_{i=1}^{m} \left(-x_{ij}\right) \ge -b_j, \quad j = 1, 2 \dots, n \right\} \left(j^{th} \text{ demand constraint}\right)$$

...(7.7)

$$\sum_{j=1}^{m} x_{ij} \ge a_i, \qquad j = 1, 2, \dots, m$$
$$\left. \sum_{i=1}^{m} \left(-x_{ij}\right) \ge -a_j, \quad j = 1, 2 \dots, m \right\} \left(i^{th} \text{ demand constraint}\right)$$

$$x_{ij} \ge 0$$

Let $\left(u_i', u_i''\right)$ be the dual variables for the ith supply constraint and $\left(v_j', v_j''\right)$ be the dual variables associated with the jth demand constraint. The dual of the given transportation problem can be written as:

$$Max \; w = \sum_{i=1}^{m} \left(u_i' - u_i''\right) a_i + \sum_{j=1}^{n} \left(v_j' - v_j''\right) b_j$$

Subject to

$$\left(u_i' - u_i''\right) + \left(v_j' - v_j''\right) \le c_{ij}$$
$$u_i', u_i'', v_j' \text{ and } v_j'' \ge 0 \; \forall i \text{ and } j$$

...(7.8)

By putting $u_i = \left(u_i' - u_i''\right)$ and $v_j = \left(v_j' - v_j''\right)$ in (7.8), we get

$$Max \; w = \sum_{i=1}^{m} u_i a_i + \sum_{j=1}^{n} v_j b_j$$

Subject to

$$u_i + v_j \le c_{ij}$$
$$u_i, v_j \text{ are unrestricted in sign } \forall \; i \text{ and } j$$

...(7.9)

From the primal-dual relationship of the transportation problem, the following observations can be made:

(i) If $x_{ij} \; \forall \; i$ and j constitute a feasible solution to the original problem, them u_i and $v_j \; \forall \; i$ and j will constitute a feasible solution to its dual problem.

(ii) If $(c_{ij} - u_i - v_j)$ denotes the value of the slack variable of the dual constraint, then $(c_{ij} - u_i - v_j) x_{ij} = 0$ by the complementary slackness theorem of duality.

Hence, we have if $x_{ij} > 0$ (which is true for basic variables) then $c_{ij} - u_i - u_j = 0$, and if $c_{ij} - u_i - u_j > 0$ (which is true for z row coefficients of non-basic variables in the primal transportation problem), then $x_{ij} = 0$.

Now to apply the MODI method, we begin with an initial basic feasible solution obtained by using any of the methods discussed in section (7.3). After obtaining the initial basic feasible solution, the following steps can be followed:

1. Compute the values of $u_i's$ for each row and $v_j's$ for each column. Since there are $(m + n)$ variables whose values have to be computed and the number of occupied cells are $(m + n - 1)$, initially to start the process, put *any one* $u_i = 0$ (u_i for the ith row containing a maximum number of allocations can also be taken as zero, initially) and use the condition $u_i + v_j = c_{ij}$ corresponding to the occupied cells in the ith row to get the value of v_j in the jth column of the given occupied cells. Now use the just obtained v_j and the condition $u_i + v_j = c_{ij}$ for all occupied cells present in the jth column to compute $u_i's$ corresponding to the rows of these occupied cells. This process is repeated again and again until all values of u_i and v_j are obtained.

2. Compute the improvement index d_{ij} for all unoccupied cells by using $d_{ij} = c_{ij} - (u_i + v_j)$. Where $i = 1, 2 ... m, j = 1, 2 ... n$.

3. If all $d_{ij} \geq 0$, then the optimum solution is reached; otherwise, select the non-basic variable or unoccupied cell (with no allocation) which has the most negative index (ties are broken arbitraily) to enter the basis.

4. By beginning with the chosen unoccupied cell, we contruct a loop via basic cells. Starting with a plus (+) sign at the new basic variable, put a minus (−) and plus (+) sign at each basic cell of the loop alternately.

5. Now, to improve the current solution further, select the "smallest" value of all basic variables appearing in the loop with a minus (−) sign. This smallest number is added to all the cells on the closed loop with a plus (+) sign and subracted from all cells on the path assigned with a minus (−) sign. In doing so at least one of the basic variables (whose allocation value becomes zero) will leave the basis (if more than one basic variable takes a zero value, the next table will have a degenerate basic feasible solution). And thus, we will have an improved basic feasible solution.

6. Repeat the previous steps untill all improvement indices computed are greater than or equal to zero.

Note: The MODI method for solving the transportation problem is also called the Modified Simplex method since it follows the same procedure as is followed by the simplex method, with minor modifications. First, the initial basic feasible solution is obtained by VAM or any other method discussed earlier and we make one non-basic variable as the basic variable and depart one basic variable from our current basic feasible solution so that the number of basic variables remains the same in the solution. By constructing the loop we decide about the leaving basic variable, and after doing some manipulations, we try to get a new basic feasible solution which gives a maximum improvement in the value of the objective function. The same procedure is followed repeatedly until no further improvement is possible, and hence we get an optimal solution.

Example 7.8. *Use the Modified distribution (MODI) method for Example 7.5.*

Solution: In the given example, the initial basic feasible solution is given as $x_{11} = 16$, $x_{12} = 6$, $x_{22} = 6$, $x_{23} = 9$, $x_{33} = 8$, and the corresponding cost of transportation is 250. Since the given problem has three supply and demand points, we have three values of $u_i's$ and $v_j's$. Initially putting $u_1 = 0$, the remaining values of u_i and v_j, $i = 1,2,3$, $j = 1,2,3$ are computed by using the relationship $u_i + v_j = c_{ij}$ for the basic cells.

market

		P	Q	R	Supply	u_i
	A	6 **16**	3 6	5 d_{13}	22	0
Warehouse	B	5 d_{21}	9 6	2 9	15	6
	C	5 d_{31}	7 d_{32}	8 8	8	12
	Demand	16	12	17	45	
	v_j	6	3	−4		

Now the next step is to compute the improvement index $d_{ij} = c_{ij} - (u_i + v_j)$ for the unoccupied cells which are written at the bottom left corner of the cell.

At the cell $(1,3)$, $d_{13} = c_{13} - (u_1 + v_3) = 5 - (0 - 4) = 1$

At the cell $(2,1)$, $d_{21} = c_{21} - (u_2 + v_1) = 5 - (6 + 6) = -7$

At the cell $(3,1)$, $d_{31} = c_{31} - (u_3 + v_1) = 5 - (12 + 6) = -13$

At the cell $(3,2)$, $d_{32} = c_{32} - (u_3 + v_2) = 7 - (12 - 3) = -8$

Since all d_{ij} are not ≥ 0, the present basic feasible solution is not the optimal basic feasible solution. Now the cell with the most negative improvement index will enter the basic. In this case, the non-basic variable corresponding to the cell $(3, 1)$ will enter the basic. Starting with the newly entered basic cell $(3, 1)$, we construct a loop via basic cells, that is, a loop is constructed such that basic cells appear at the corner points of the loop. Begining with the positive sign at the new cell, alternately put a minus $(-)$ and plus $(+)$ sign at each basic cell of the loop.

market

Warehouse		P	Q	R	Supply	u_i
	A	6 16 ----→6 – ▲　+	3 　+	5 1	22	0
	B	5	5 6 ----→9	9 　+	15	6
		–7		2		
	C	+◄----	5 7	– 8 8	8	12
		–13	–8			
	Demand	16	12	17	45	
	v_j	6	3	–4		

To improve the current solution further, select the "smallest" value of basic variables in the loop containing a minus $(-)$ sign. In this case, 6 is the minimum value among all basic variables containing a $(-)$ sign. This number is added to all the cells on the closed loop with a plus $(+)$ sign and subtracted from all cells on the path assigned with a minus $(-)$ sign. In doing so, the variable in the cell (2, 2) will become a non-basic variable in the next iteration, and thus we will have an improved basic feasible solution as follows:

market

Warehouse		P	Q	R	Supply
	A	6 10	3 12	5	22
	B	5	9	2 15	15
	C	5 6	7	8 2	8
	Demand	16	12	17	45

Now the new improved solution is $x_{11} = 10$, $x_{12} = 12$, $x_{23} = 15$, $x_{31} = 6$, and $x_{33} = 2$, and the corresponding total cost of transportation is $z = 172$. By repeating the previous steps again, we get the following:

Warehouse		P	Q	R	Supply	u_i
	A	-6 10	3 12	$+5$ -4	22	0
	B	5 4	9 5	2 15	15	-7
	C	$+$ 5 6	7 5	$-$ 8 2	8	-1
	Demand	16	12	17	45	
	v_j	6	3	9		

Further we have:

		market P	Q	R	Supply	u_i
	A	6 8	3 12	5 2	22	0
Warehouse	B	5 2	9 9	2 15	15	− 3
	C	5 8	7 5	8 4	8	− 1
	Demand	16	12	17	45	
	v_j	6	3	5		

Now since all the improvement indices are ≥ 0, the optimum solution has been reached.

Here, $x_{11} = 8, x_{12} = 12, x_{13} = 2, x_{23} = 15,$ and $x_{31} = 8$ is the optimal basic feasible solution, and the corresponding total minimum cost of transportation is $z = 164$.

7.6 Solution of Transportation Problems in Excel

A transportation problem can easily be solved using the Solver add-in of MS-Excel. We use Example 7.1 to explain the steps involved.

1. Copy the given problem on the Excel sheet.

C	D	E	F	G	H	I
	D1	D2	D3	D4	D5	Supply
O1	4	4	5	6	4	20
O2	7	8	5	9	4	35
O3	5	4	11	6	9	40
Demand	25	10	30	15	15	95

Fig. 7.2

2. Create a solution space on the same Excel sheet as follows:

C	D1	D2	D3	D4	D5	Supply
O1	4	4	5	6	4	20
O2	7	8	5	9	4	35
O3	5	4	11	6	9	40
Demand	25	10	30	15	15	95

C	D1	D2	D3	D4	D5	Supply
O1						20
O2		Space for Solution				35
O3						40
Demand	25	10	30	15	15	95

Fig. 7.3

3. Insert a blank column and row alongside the space provided for solution (x_{ij}) to compute the row and column sum, as shown in the following figure:

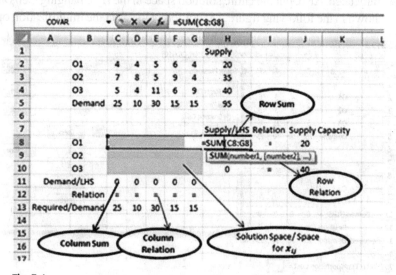

Fig. 7.4

4. Create a space for computing the value of the total minimum cost of transportation. The **SUMPRODUCT** function will be used to multiply the given unit cost and solution space.

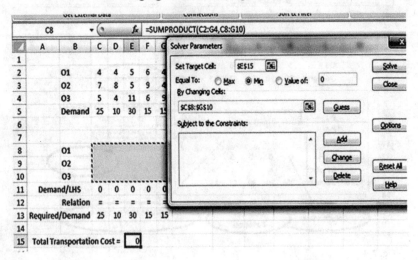

Fig. 7.5

5. Now go to the Solver. Put the Total Transportation cost cell address in the Set Target cell. Also, put the entire solution space in the By Changing Cells area, as shown in the following figure. Do not forget to select the minimization option.

Fig. 7.6

6. Taking the column and row sums as the LHS and the given supply and demand as the RHS of the constraints, we have a total of $3 + 3 = 6$ constraints. Enter this information as follows:

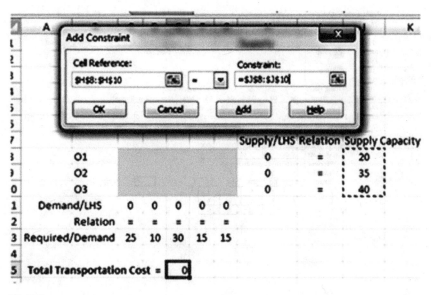

Fig. 7.7

7. Select the Simplex LP option from the dropdown. Then click solve.

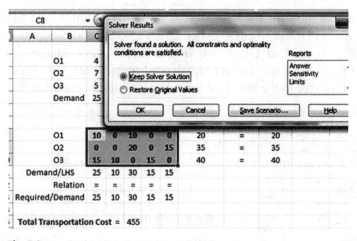

Fig. 7.8

8. The previous solution is in agreement with the optimum solution obtained using the stepping-stone method. And below the following is the calculation of the same using the Jensen add-in.

Transportation Model			Name: Trans_1			Objective TerrSolver: Jensen Network				Ph. 1 Iter. 18
TRUE			Type: Trans 'ansportation:		455	Type: Linear				Total Iter. 27
FALSE	● Change		Goal: Min		Suppliers:	0 Sens.: No				Comp. Time 00:00
TRUE			Cost: 455		Demanders:	0 nteger: No				Status Optimal
FALSE	● Solve					Side: No				
FALSE										
100	● Vary									
100										

		Trans. Flows	1	2	3	4	5	Supply Data			
0		Name	D1	D2	D3	D4	D5	Min.	Max.	Cost	Shipped
60											
FALSE	1	S1	10	0	10	0	0	20	10000	0	20
FALSE	2	S2	0	0	20	0	15	35	10000	0	35
FALSE	3	S3	15	10	0	15	0	40	10000	0	40
Demand		Min.:	25	10	30	15	15				
Data		Max:	10000	10000	10000	10000	10000				
		Cost:	0	0	0	0	0				
		Received:	25	10	30	15	15				
		Trans. Cost	1	2	3	4	5				
		Name	D1	D2	D3	D4	D5				
	1	S1	4	4	5	6	4				
	2	S2	7	8	5	9	4				
	3	S3	5	4	11	6	9				

Fig. 7.9

7.7. Some Special Cases in Transportation Problems

7.7.1. Unbalanced Transportation Problems

We have already proved that $\sum_{i=1}^{m} a_i = \sum_{j=1}^{n} b_j$ is the necessary and sufficient condition for the existence of a feasible solution to a transportation problem, and the problem is called a balanced transportation problem if this condition is satisfied. Otherwise, if total supply is not equal to total demand, then the problem is called an unbalanced transportation problem and it has to be converted to the balanced form to get an initial basic feasible solution.

Here two cases arise:

(i) $\sum_{i=1}^{m} a_i > \sum_{j=1}^{n} b_j$ i.e. Total supply > total demand

In this case, the constraints take the following forms:

$$\sum_{i=1}^{m} x_{ij} = b_j \text{ and } \sum_{j=1}^{n} x_{ij} < a_i, i = 1, 2 \ldots m, j = 1, 2, \ldots n$$

And $\left(\sum_{i=1}^{m} a_i - \sum_{j=1}^{n} b_j \right)$ is the excess number of units available.

Hence, in order to get a balanced transportation problem, here we will add a dummy destination with the demand equal to excess availability, that is,

$\left(\sum_{i=1}^{m} a_i - \sum_{j=1}^{n} b_j\right)$ and the unit cost of transportation from any source to this dummy destination will be considered zero.

(ii) $\sum_{i=1}^{m} a_i < \sum_{j=1}^{n} b_j$ Total supply < total demand.

In this case, the constraints take the following form-

$$\sum_{i=1}^{m} x_{ij} \le b_j \text{ and } \sum_{j=1}^{n} x_{ij} = a_i, i = 1, 2 \dots m, j = 1, 2, \dots n$$

And $\left(\sum_{j=1}^{n} b_j - \sum_{i=1}^{m} a_i\right)$ is the excess demand.

Hence, in order to get a balanced transportation problem, here we will add a dummy source with the availability equal to excess demand, that is, $\left(\sum_{j=1}^{n} b_j \sum_{i=1}^{m} a_i\right)$, and the unit cost of transportation from this dummy source to any destination will be considered to be zero.

Once we convert the problem into a balanced one, it will be solved in the same manner as done earlier.

Example 7.9. *Convert the following unbalanced transportation problem into a balanced one.*

	D1	D2	D3	D4	Supply
O_1	1	3	5	2	100
O_2	2	11	7	5	150
O_3	6	4	6	15	200
Demand	125	110	160	155	

Solution: The total demand of the given problem is **550**, whereas the total supply is **450**, that is, $\left(\sum_{j=1}^{n} b_j \sum_{i=1}^{m} a_i\right)$; therefore, we introduced a dummy row (O_4) having all unit transportation costs equal to zero and a row supply equal to $550 - 450 = 100$. The following is the balanced transportation table obtained:

	D1	D2	D3	D4	Supply
O_1	1	3	5	2	100
O_2	2	11	7	5	150
O_3	6	4	6	15	200
O_4	0	0	0	0	100
Demand	125	110	160	155	**550**

Example 7.10. *Give the LPP formulation of the transportation problem given in Example 7.7. Also solve it using Excel Solver.*

Solution: The total demand of the given problem is **550**, whereas the total supply is **450**, that is, $\left(\sum_{j=1}^{n} b_j > \sum_{i=1}^{m} a_i\right)$ so it's an unbalanced transportation problem. Let x_{ij} ($i = 1, 2, 3; j = 1, 2, 3, 4$) denote the amount of units transported from the ith source to the jth destination. Then the LPP formulation of the given problem is as follows:

$$Min \ z = x_{11} + 3x_{12} + 5x_{13} + 2x_{14} + 2x_{21} + 11x_{22} + 7x_{23} + 5x_{24}$$
$$+ 6x_{31} + 4x_{32} + 6x_{33} + 15x_{34}$$

Subject to

$$\left.\begin{array}{l} x_{11} + x_{12} + x_{13} + x_{14} = 100 \\ x_{21} + x_{22} + x_{23} + x_{24} = 150 \\ x_{31} + x_{32} + x_{33} + x_{34} = 200 \end{array}\right\} \text{supply constraints}$$

$$\left.\begin{array}{l} x_{11} + x_{21} + x_{31} \leq 125 \\ x_{12} + x_{22} + x_{32} \leq 110 \\ x_{13} + x_{23} + x_{33} \leq 160 \\ x_{14} + x_{24} + x_{34} \leq 155 \end{array}\right\} \text{Demand constraints}$$

$$x_{ij} \geq 0 \ (i = 1, 2, 3; j = 1, 2, 3, 4)$$

The following is the screen shot of the Excel Solver calculation:

	D1	D2	D3	D4	Supply
O1	1	3	5	2	100
O2	2	11	7	5	150
O3	6	4	6	15	200
Demand	125	110	160	155	

	D1	D2	D3	D4	Supply/LHS	Relation	Supply Capacity
O1	0	0	0	100	100	=	100
O2	125	0	0	25	150	=.	150
O3	0	110	90	0	200	=	200
Demand/LHS	125	110	90	125			
Relation	≤	≤	≤	≤			
Required/Demand	125	110	160	155			

Total Transportation Cost = 1555

Fig. 7.10

As per the previous calculation our optimal basic feasible solution is $x_{14} = 100$, $x_{21} = 125$, $x_{24} = 25$, $x_{32} = 110$, and $x_{33} = 90$. The corresponding minimum cost of transportation is 1555. Since the total number of basic variables having a positive value is 5, which is less than $(3 + 4 - 1)$, the current optimal basic feasible solution is a degenerate one.

7.7.2. Restricted Entry

Restricted Entry: Sometimes the transportation from a given source to another destination is not possible (it is restricted or not allowed for some reason); then in such a situation we assign infinity (∞) or a very large number (cost) at that cell (i, j) and follow the previously discussed methods to find a solution to the given problem. While solving such problems using Excel Solver, one should replace infinity (∞) by a large positive number.

7.7.3. Maximization Problems

In the transportation problem, generally we minimize the total cost of transportation given the unit cost of transportation to the various locations from the available supply points. However, we may also have problems where the cell entries in a transportation table denote unit profits instead of unit costs, and hence the objective of these problems will be to maximize the total profits. In such situations, the problem cannot be solved using the previously discussed method. To solve a maximization type problem, the solution method discussed earlier cannot be applied as it is, and hence it require some changes to be made. The given problem needs to be converted into an equivalent minimization type problem. This is done by subtracting all the elements of the matrix from the largest element of the matrix. The resulting matrix is called the **opportunity loss matrix.** The cell entries of the opportunity loss matrix denote the loss of opportunity for not selecting the best option in the matrix (largest element of the matrix). Now the procedure described earlier for solving the minimization problem can be applied, and the solution can be obtained. Finally, to obtain the maximum profit, the solution obtained here will be multiplied by the corresponding cell entries of the original matrix.

Example 7.11. *A company distributes its product from three plants to four different warehouses. The monthly supplies and demands along with per-unit profit for transportation are given in the following table. Find the initial basic feasible solution using the VAM of the transportation problem so as to maximize the total profit.*

To warehouses					
From Plant	1	2	3	4	Supply
1	5	12	10	4	80
2	6	9	2	15	50
3	8	11	18	7	70
Demand	40	50	60	50	

Solution: The given problem is a maximization problem, and VAM as discussed earlier cannot be applied here in the same way. To use VAM, we need to convert this table into an opportunity loss table by subtracting all the elements of the table from 18, which is the largest per-unit profit in the table. Hence, the opportunity loss table is as follows:

To warehouses					
From Plant	1	2	3	4	**Supply**
1	13	6	8	14	80
2	12	9	16	3	50
3	10	7	0	11	70
Demand	40	50	60	50	

Now we use VAM for this table:

To warehouses						Penalty		
From Plant	1	2	3	4	**Supply**	**I**	**II**	**III**
1	13 30	6 50	8	14	80 30	2	7	7
2	12	9	16	3 50	50	6	6	
3	10 10	7	0 60	11	70 10	7	3	3
Demand	40 10	50	60	50				
I	2	1	8	8				
II	2	1		8				
III	3	1						

Placing the previous allocation in the original table, we get:

To warehouses					
From Plant	1	2	3	4	**Supply**
1	5	12	10	4	80
	30	**50**			
2	6	9	2	15	50
			50		
3	8	11	18	7	70
	10		**60**		
Demand	40	50	60	50	

So, our initial basic feasible solution obtained using VAM is $x_{11} = 30$, $x_{12} = 50$, $x_{24} = 50$, $x_{31} = 10$, and $x_{33} = 60$. And the corresponding total profit is $(5 * 30 + 12 * 50 + 15 * 50 + 8 * 10 + 18 * 60) = 2660$.

Use of Excel Solver for solving the previous maximization transportation problem

When using Solver to solve the previous problem, we don't need to convert the given transportation table into an opportunity loss table. Here the problem will be solved as a maximization problem instead of a minimization problem by selecting the maximization option in place of minimization as shown in the following figure:

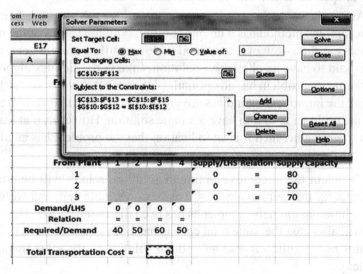

Fig. 7.11

Now by clicking the solve option after the necessary settings, we have following optimum solution:

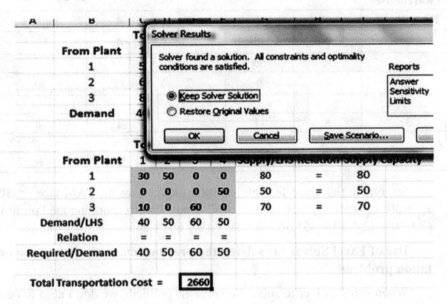

Total Transportation Cost = 2660

Fig. 7.12

7.7.4. Multiple Solutions in Transportation Problems

As we are aware that any linear programming problem while being solved by the simplex method has a unique solution if all the values of the non-basic variables in the z-row of the optimal table are different from zero, the problem is said to have an alternative solution. Like in the simplex algorithm, the transportation problem has an optimal solution if all $d_{ij} = c_{ij} - (u_i + v_j) \geq 0$, that is, if all the improvement indices are greater than or equal to zero. If all $d_{ij} > 0$, then the problem is said to have a unique solution. However, if at least one d_{ij} takes a zero value, then it is an indication that the problem has an alternative solution.

To obtain the alternative optimum solution, a loop will be constructed starting with the cell having $d_{ij} = 0$ as basic variable cell, and we will get a new solution in the same manner as has been discussed earlier. The new solution will also give the same total cost as has been obtained earlier, indicating that this new solution is also an optimal solution to the same transportation problem.

7.7.5. Degeneracy in Transportation Problems

If the number of basic variables in a basic feasible solution to a transportation problem is less than ($m + n - 1$), then the solution is said to be a degenerate basic feasible solution. In Example 7.8, the number of basic variables in the initial basic feasible solution obtained using VAM was 5, which is less than $(3 + 4 - 1) = 6$. And hence, this solution will be a degenerate basic feasible solution, since here we will generate one basic variable (to make it equal to ($m + n - 1 = 6$) with the value zero. Degeneracy in a transportation problem can arise in the following situations:

(i) The initial basic feasible solution has less than $m + n - 1$ basic variables (as in the case of Example 7.8).

(ii) While moving towards the optimality, when two or more basic variables leave the basis simultaneously, the next basic feasible solution will have less than $m + n - 1$ basic variables, thus making the solution degenerate.

The difficulty with degeneracy is, the optimality test of the problem cannot be completed. The stepping-stone method fails because for some of the non-basic cells, the closed loops cannot be formed, while in the MODI method, we will not be able to find all the values of u_i and v_j so as to complete the optimality test.

To resolve degeneracy, we assign a very small positive number (close to zero) denoted by δ to one or more non-basic cells (as required) so as to make the total number of basic variables equal to ($m + n - 1$). The number δ theoretically is non-zero, and mathematically it is considered as good as zero; that is, for any constant 'a' we have: $\delta + a = a$; $\delta + \delta = \delta \times \delta = 0$; $a - \delta = a$; $\delta \times a = 0$. The allocation of δ should not be done randomly; instead, this should be placed at independent location(s) (**the location is called independent if, starting with the given location, we try to construct a loop by having basic variables at the corner of the loop but we are not able to do so**). If this is not done, then we will not be able to find all the values of u_i and v_j, and hence the degeneracy will still be unresolved. Out of all available independent non-basic cells, we should prefer the one with the minimum cost to put into δ.

Example 7.12. *Find the optimum solution to the following cost-minimizing transportation problem.*

From	To	D1	D2	D3	Supply
	S1	10	9	5	100
From	S2	5	10	11	110
	S3	13	5	7	120
	Demand	90	120	120	330

Solution: Since the total demand is equal to the total supply, the given problem is a balanced one. The following is the initial basic feasible solution of the problem obtained using VAM:

From \ To	D1	D2	D3	Supply
S1	15	9	5 **80**	80
S2	5 **90**	10	11 **20**	110
S3	13	10 **120**	12	120
Demand	90	120	100	310

Here since the total number of basic variables is **4,** which is léss than $m + n - 1 = 3 + 3 - 1 = 5$, the previous basic feasible solution is a degenerate basic feasible solution. Here except for cell (1, 1), the remaining cells (1, 2), (2, 2), (3, 1), and (3, 3) are independent. We now introduce a very small positive number in cell (1, 2) (which has the minimum per unit cost among the independent cells) and consider this as one of the basic cells. And then we apply the **MODI** method to check the optimality of the present solution.

From \ To	D1	D2	D3	Supply	u_i
S1	15 16	9 $-\delta$	5 ▶80+	80	0
S2	5 **90**	10 + −5	11 20−	110	6
S3	13 13	10 **120**	12 6	120	1
Demand	90	120	100	310	
v_j	−1	9	5		

Since all the improvement indexes are not greater than or equal to zero, the optimality condition is yet to be fulfilled. So, we construct a loop as shown in the previous table, and a rearrangement of allocation is done as discussed before.

From \ To	D1	D2	D3	Supply	u_i
S1	15 16	9 4	5 **80**	80	0
S2	5 **90**	9 δ	11 **20**	110	6
S3	13 9	10 **120**	12 2	120	5
Demand	90	120	100	310	
v_j	−1	5	5		

Here all the improvement indices are now greater than zero. The optimality condition is fulfilled. The optimal basic feasible solution is $x_{13} = 80$, $x_{21} = 90$, $x_{23} = 20$, and $x_{32} = 120$, and the corresponding minimum cost of transportation is $5 \times 80 + 5 \times 90 + 11 \times 20 + 10 \times 120 = 2270$.

The following is the MS-Excel calculation of the previous problem.

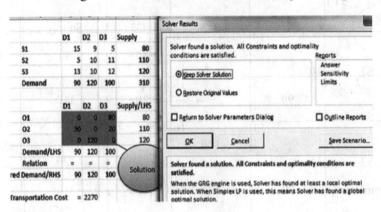

Fig. 7.13

7.8. Transshipment Problems

In the transportation problem, we have seen that the items are directly shipped from a particular source to a particular destination in such a way that the total cost of transportation is at a minimum. The movement of the items among sources or among destinations or the movement of the items through some intermediate points is not allowed in a transportation problem. The shipping problems with any or all of these characteristics are called transshipment problems. And hence, a transshipment problem is an extension of the transportation problem. The optimal solution to a transshipment problem can be found by using the methods discussed in the transportation problem. In this book we discussed two types of transshipment problems:

(i) Transshipment problems with some intermediate points between the supply and demand points; and

(ii) Transshipment problems where movement of items are allowed between one supply point and another supply point and also between one demand point and another demand point without having any separate additional transshipment points.

The following figure is an example of a transshipment problem with 4 supply and 3 demand points and two intermediate points where S_i ($1 = 1, 4$) denotes the supply available at i^{th} supply points, T_j ($j = 1, 2$) denotes the jth

intermediate points, and D_k ($k = 1, 2, 3$) denotes the demand at the kth destination point. Here the movement of items among the intermediate points is restricted.

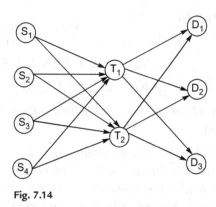

Fig. 7.14

We denote x_{ij} ($i = 1, 2, 3, 4$ and $j = 1, 2$) the amount of units shipped from the ith supply point to the jth intermediate point and x'_{jk} ($j = 1, 2$ and $k = 1, 2, 3$) denotes the amount of units shipped from the jth intermediate point to the kth demand point. Also let cij, ($i = 1, 2, 3, 4$ and $j = 1, 2$) denote the unit cost of shipping from the ith supply point to the jth intermediate point and c'_{jk}, ($j = 1, 2$ and $k = 1, 2, 3$) the unit cost of shipping from the jth transshipment point to the kth demand point.

The linear programming formulation of the previous problem can be written as:

$$Min\ Z = c_{11}x_{11} + c_{12}x_{12} + c_{21}x_{21} + c_{22}x_{22} + c_{31}x_{31} + c_{32}x_{32}$$
$$+ c_{41}x_{41} + c_{42}x_{42} + + c'_{11}x'_{11} + c'_{12}x'_{12} + c'_{13}x'_{13}$$
$$+ c'_{21}x'_{21} + c'_{22}x'_{22} + c'_{23}x'_{23}$$

Subject to

$$\left. \begin{array}{l} x_{11} + x_{12} = S_1 \\ x_{21} + x_{22} = S_2 \\ x_{31} + x_{32} = S_3 \\ x_{41} + x_{42} = S_4 \end{array} \right] \text{Supply constraints}$$

$$\left.\begin{array}{l} x'_{11} + x'_{12} = D_1 \\ x'_{12} + x'_{22} = D_2 \\ x'_{13} + x'_{23} = D_3 \end{array}\right] \text{Demand constraints}$$

$$\left.\begin{array}{l} x_{11} + x_{21} + x_{31} + x_{41} = x'_{11} + x'_{12} + x'_{13} \\ x_{12} + x_{22} + x_{32} + x_{42} = x'_{21} + x'_{22} + x'_{23} \end{array}\right] \text{Transshipment point constraints}$$

$$x_{ij} \geq 0 \ (i = 1, 2, 3; j = 1, 2, 3, 4)$$

and $\qquad x'_{jk} \geq 0 \ (j = 1, 2; k = 1, 2, 3)$

Example 7.13. (*Example with transshipment points between the supply and demand points*)

A company has three factory locations where items are produced and are finally shipped to three locations (called demand points) through two transshipment points. It is also given that items cannot be directly transported to the demand points and movements between the transshipment points are also restricted. The following tables show the total items produced at the three locations, the different requirements at the three demand points, and the unit cost of shipment from the factories to the transshipment points and from the transshipment points to the various demand points.

Factory	Transshipment Points		Items Available
	T_1	T_2	
S_1	21	19	200
S_2	18	16	150
S_3	15	17	200

Transshipment Points	Units Demanded		
	D_1	D_2	D_3
T_1	12	16	14
T_2	14	19	17
Items Required	190	200	160

Solve the previous transshipment problem to find the optimum transportation schedule from the factories to the demand points via the transshipment points.

Solution: The total available items in the three supply points are 550 and the total requirements at the three demand points are 550. Since total units available are equal to the total demand, the given problem is a balance transshipment problem.

As the items cannot be shipped directly from the supply points to the demand points, the per unit direct cost of shipping from the three supply points to the three demand points will be put equal to a very large number, for example, infinity.

If x_{ij} ($i = 1, 2, 3$ and $j = 1, 2$) denotes the amount of units shipped from the ith supply point to the jth transshipment point and x'_{jk} ($j = 1, 2, 3$ and $k = 1, 2, 3$) denotes the amount of units shipped from the jth transshipment point to the kth demand point, then the LPP formulation of the previous problem is given as follows:

$$Min\ z = 21x_{11} + 19x_{12} + 18x_{21} + 16x_{22} + 15x_{31} + 17x_{32}$$
$$+ 12x'_{11} + 16x'_{12} + 14\,x'_{13} + 14\,x'_{21} + 19\,x'_{22} + 17\,x'_{23}$$

Subject to

$$\left.\begin{array}{l} x_{11} + x_{12} = 200 \\ x_{21} + x_{22} = 150 \\ x_{31} + x_{32} = 200 \end{array}\right\} \text{Supply constraints}$$

$$\left.\begin{array}{l} x'_{11} + x'_{21} = 190 \\ x'_{12} + x'_{22} = 200 \\ x'_{13} + x'_{22} = 160 \end{array}\right\} \text{Demand constraints}$$

$$\left.\begin{array}{l} x_{11} + x_{21} + x_{31} = x'_{11} + x'_{12} + x'_{13} \\ x_{12} + x_{22} + x_{32} = x'_{21} + x'_{22} + x'_{23} \end{array}\right\} \text{Transshipment point constraints}$$

$$x_{ij} \geq 0\ (i = 1, 2, 3; j = 1, 2)$$

and $\qquad\qquad x'_{ji} \geq 0\ (i = 1, 2, 3; j = 1, 2)$

To start using VAM to find the initial basic feasible solution to the given problem, we consider the two transshipment points as additional supply and demand points, with availability and requirement as 550 (sum of the given availability/demand) units each, and of course the movement between these transshipment points still restricted, which is shown in the table by infinity. Now the given problem can be rewritten as:

	D_1	D_2	D_3	T_1	T_2	Items Available
S1	∞	∞	∞	21	19	200
S2	∞	∞	∞	18	16	150
S3	∞	∞	∞	15	17	200
T_1	12	16	14	0	∞	550
T_2	14	19	17	∞	0	550
Items Required	190	200	160	550	550	

Now the transshipment table has a total of five rows and five columns. The penalties are calculated for the rows and columns as shown in the following:

	D_1	D_2	D_3	T_1	T_2	Items Available	Penalty
S_1	∞	∞	∞	21	19	200	2
S_2	∞	∞	∞	18	16	150	2
S_3	∞	∞	∞	15	17	200	2
T_1	12	16	14	0	∞	550	12
T_2	14	19	17	∞	0 550	550	14
Items Required	190	200	160	550	550		
Penalty	2	3	3	15	16		

The highest penalty 16 corresponds to the fifth column, and hence we allocate the maximum possible units to the cell (T_2, T_2) where the unit cost is minimum and the following is the reduced matrix after the necessary deletion:

	D_1	D_2	D_3	T_1	Items Available	Penalty
S_1	∞	∞	∞	21	200	∞
S_2	∞	∞	∞	18	150	∞
S_3	∞	∞	∞	15	200	∞

T_1	12 **190**	16	14	0	~~550~~ 360	12
Items Required	~~190~~	200	160	550		
Penalty	∞	∞	∞	15		

Now the highest penalty in the previous table is ∞, which occurs at six different places. The cell (T_1, D_1) has a minimum cost among these, and hence we allocate the maximum possible amount in this cell. The following is the reduced matrix.

	D_2	D_3	T_1	Items Available	Penalty
S_1	∞	∞	21	200	∞
S_2	∞	∞	18	150	∞
S_3	∞	∞	15	200	∞
T_1	16	14 **160**	0	~~360~~ 200	16
Items Required	200	160	550		
Penalty	∞	∞	15		

Maximum penalty is again ∞, which occurs at five different locations, 14 at (T_1, D_3) has the minimum cost among these rows and columns, and hence the maximum possible allocation is done at this cell. The reduced matrix is shown as follows:

	D_2	T_1	Items Available	Penalty
S_1	∞	21	200	∞
S_2	∞	18	150	∞
S_3	∞	15 **200**	~~200~~	∞
T_1	16	0	200	16
Items Required	200	~~550~~ 350		
Penalty	∞	15		

The previous steps are repeated again and again as in the following, until we have the final initial basic feasible solution:

	D_2	T_1	Items Available	Penalty
S_1	∞	21	200	∞
S_2	∞	18	150	∞
T_1	16 **200**	0	~~200~~	16
Items Required	~~200~~	350		
Penalty	∞	18		

	T1	Items Available	Penalty
S_1	21 **200**	200	∞
S_2	18 **150**	150	∞
Items Required	350		
Penalty	3		

Thus, the following is the initial basic feasible solution to the given problem obtained using VAM:

	D_1	D_2	D_3	T_1	T_2	Items Available
S_1	∞	∞	∞	21 **200**	19	200
S_2	∞	∞	∞	18 **150**	16	150
S_3	∞	∞	∞	15 **200**	17	200
T_1	12 **190**	16 **200**	14 **160**	0	∞	550
T_2	14	19	17	∞	0 **550**	~~550~~
Items Required	190	200	160	550	550	

The total cost of transshipment for the present allocation = 21 * 200 + 18 * 150 + 15 * 200 + 12 * 190 + 16 * 200 + 14 * 160 = 17620.

Since the total number of allocation is 7, which is less than $(5 + 5 - 1)$, the previous initial basic feasible solution is a degenerate basic feasible solution. To apply the MODI method, we introduce a very small number δ at $(T_1 T_1)$ and $(T_2 D_2)$, and then we have:

	D_1	D_2	D_3	T_1	T_2	Items Available	u
S_1	∞	∞	∞	21 **200**	19	200	0
S_2	∞	∞	∞	18 **150**	16	150	−3
S_3	∞	∞	∞	15 **200**	17	200	−6
T_1	12 **190**	16 **200**	14 **160**	0 **δ**	∞	550	−21
T_2	14 **δ**	19	17	∞	0 **550**	550	−19
Items Required	190	200	160	550	550		
v	33	37	35	21	19		

The following are the improvement indexes for the non-basic cells:

$$(S_1, T_2) = 19 - (0 + 19) = 0$$
$$(S_2, T_2) = 16 - (-3 + 19) = 0$$
$$(S_3, T_2) = 17 - (-6 + 19) = 4$$
$$(T_2, D_2) = 19 - (-19 + 37) = 37$$
$$(T_2, D_3) = 17 - (-19 + 35) = 35$$

Since all the improvement indexes are ≥ 0, the current basic feasible solution is the optimal basic feasible solution.

The following is the pictorial representation of the optimal solution:

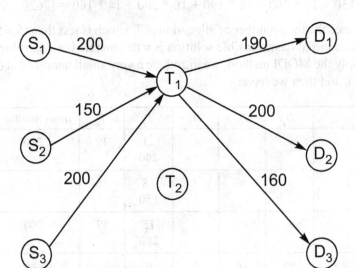

Fig. 7.15

Note: Since the improvement index corresponding to (S1, T2) and (S2, T2) are equal to zero, the present problem has more than one optimum solution. If we allow either of these two cells to enter the basis, we can find an alternative optimum solution to the given problem.

Use of Excel Solver for solving the previous transshipment problem:

The information on the Excel sheet will be entered as it was entered in the transportation problem. Here we have three types of constraints, namely, supply constraints, demand constraints, and transshipment constraints, as is seen from the LPP formulation of the given problem. Since infinity is not recognized by Excel, replace it with 1,000 (large positive number), which can be considered as large enough in the given context. The final output solution of the problem using Excel Solver is as follows:

Using the normal transportation approach:

B	C	D	E	F	G	H	I	J
	D_1	D_2	D_3	T_1	T_2	Items Available		
S_1	1000	1000	1000	21	19	200		
S_2	1000	1000	1000	18	16	150		
S_3	1000	1000	1000	15	17	200		
T_1	12	16	14	0	1000	550		
T_2	14	19	17	1000	0	550		
Items Required	190	200	160	550	550			

	D_1	D_2	D_3	T_1	T_2	LHS	Relation	RHS
S_1	0	0	0	200	0	200	=	200
S_2	0	0	0	150	0	150	=	150
S_3	0	0	0	200	0	200	=	200
T_1	190	200	160	0	0	550	=	550
T_2	0	0	0	0	550	550	=	550
LHS	190	200	160	550	550			
Relation	=	=	=	=	=			
RHS	190	200	160	550	550			

Total minimum cost of transshipment=   17620

Fig. 7.16

Using the LPP formulation

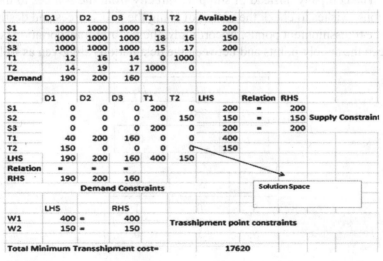

	D1	D2	D3	T1	T2	Available	
S1	1000	1000	1000	21	19	200	
S2	1000	1000	1000	18	16	150	
S3	1000	1000	1000	15	17	200	
T1	12	16	14	0	1000		
T2	14	19	17	1000	0		
Demand	190	200	160				

	D1	D2	D3	T1	T2	LHS	Relation	RHS	
S1	0	0	0	200	0	200	=	200	
S2	0	0	0	0	150	150	=	150	Supply Constraint
S3	0	0	0	200	0	200	=	200	
T1	40	200	160	0	0	400			
T2	150	0	0	0	0	150			
LHS	190	200	160	400	150				
Relation	=	=	=						
RHS	190	200	160						

Demand Constraints

Solution Space

	LHS		RHS	
W1	400	=	400	Trasshipment point constraints
W2	150	=	150	

Total Minimum Transshipment cost=   17620

Fig. 7.17

The previous calculation is done using the LPP formulation of the given problem, and it represents an alternative optimum feasible solution for the given problem. It can be clearly seen that only the necessary movements at the transshipment points are used in the transshipment constraints.

Example 7.14. (*Problem where the movement of the items is allowed between one supply point and another supply point and also between one demand point and another demand point without having any separate additional transshipment points*)

A company has a transshipment problem with two supply points S_1 and S_2 from which items are to be shipped to the two different destinations D_1 and D_2. The unit cost of transportation between supply and demand points and the availabilities and requirements at various sources and destinations are given in the following table:

	D_1	D_2	Availability
S_1	5	6	30
S_2	4	4	10
Requirements	15	25	

The company, instead of shipping directly from the sources to the destinations, wishes to check the option of using the two sources and destinations as the transshipment points. Accordingly, the per unit transportation costs between the supply points and between the destinations are given as follows:

	D_1	D_2
D_1	0	2
D_2	2	0

	S_1	S_2
S_1	0	1
S_2	1	0

Solve the previous transshipment problem to find the optimum transportation schedule from the supply to the demand points.

Solution: Here total availability is equal to the total requirements, and hence the given transshipment problem is a balanced problem. Since each supply and destination point can act as a transshipment point, the given problem can be represented by a (4 × 4) cost matrix as follows:

	S_1	S_2	D_1	D_2	Availability
S1	0	1	5	6	30
S2	1	0	4	4	10
D1	5	4	0	2	
D2	6	4	2	0	
Requirement			15	25	

To solve such a problem, we first add 40 units (the total number of units which are to be transported from the supply point to the destinations) to all the supply point availabilities and destination points demands. The revised table is shown as follows:

	S_1	S_2	D_1	D_2	
S_1	0	1	5	6	70
S_2	1	0	4	4	50
D_1	5	4	0	2	40
D_2	6	4	2	0	40
	40	40	55	65	

We now can solve this modified problem as a normal transportation problem. Finding the initial basic feasible solution using VAM and testing its optimality using the MODI method, the following is the optimal solution to the given problem (steps are left for readers to verify):

	S_1	S_2	D_1	D_2	
S_1	0 40	1 15	5 15	6	70
S_2	1	0 25	4	4 25	50
D_1	5	4	0 40	2	40

D_2	6	4	2	0 **40**	40
	40	40	55	65	

Leaving the allocations in the diagonal cells, the following are the optimal allocations to the given problem:

$$S_2 \rightarrow S_2 = 15; \qquad S_1 \rightarrow D_1 = 15 \quad \text{and} \qquad S_2 \rightarrow D_2 = 25$$

The total minimum cost of transshipment is 190, and the following is the network representation of the previous solution.

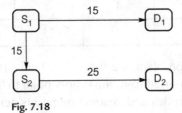

Fig. 7.18

The following is the calculation of the previous problem done using Excel Solver:

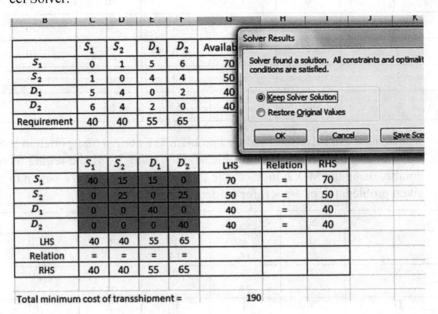

Fig. 7.19

Exercises

1. What is a transportation problem? Give the LPP formulation of a general transportation problem and give the matrix representation of the same.

2. How do the decision variables in a transportation problem differ from the decision variables in a normal linear programming problem?

3. Show that the necessary and sufficient condition for the existence of a feasible solution to a transportation problem is:

$$\sum_{i=1}^{m} a_i = \sum_{j=1}^{m} b_j$$

Where a_i ($i = 1, 2, ..., m$) is the availability of the ith supply point and b_j ($j = 1, 2,...,n$) is the requirement of the jth demand point.

4. Define a loop in a transportation problem and give its importance while solving a transportation problem.

5. Explain the various methods for finding the initial basic feasible solution to a transportation problem.

6. What is a balanced transportation problem? How is this important in the solution procedure?

7. Explain an unbalanced transportation problem.

8. A firm has three plants which produce items of similar types, and the monthly production capacities of these plants are 1,900, 2,099, and 1,200. These items are to be distributed to four different locations. Sales forecasts indicate that monthly deliveries will be 999, 1,500, 1,700, and 1.000 to locations 1 to 4, respectively. The transportation cost of shipping a unit from a plant to a location is given as follows:

Location	1	2	3	4
Plant 1	15	20	11	13
Plant 2	25	14	10	17
Plant 3	9	13	12	16

Formulate a linear programming model for the previous transportation problem. Use the North-West Corner Rule and the least-cost method to find the initial basic feasible solution to the given transportation problem. Does the use of the least-cost method give a better solution than the North-West rule?

9. What are the two methods for testing the optimality of a given transportation problem? Explain them. Which of the two methods would you prefer and why?

10. Explain the connection between the MODI method for solving a transportation problem and the complementary slackness theorem of duality.

11. What is degeneracy and how does it arise in a transportation problem? What problem does this create in the MODI method for testing optimality? Explain how you overcome it.

12. Explain how to solve a profit maximization transportation problem.

13. How do you identify the existence of an alternative solution to a transportation problem, and how will you find it?

14. Use VAM to find the initial basic feasible solution to Question 8 and check for optimality. If it is not optimal, find the optimal solution.

15. Show that the optimum solution to a transportation problem remains unchanged when a fixed constant is added to every element in the given cost matrix.

16. Solve the following transportation problems, whose cost matrix is given as follows (use VAM to find the initial basic feasible solution):

(i)

	Destination				Availability
Origin	1	2	3	4	
O1	11	9	21	16	400
O2	15	18	11	19	200
O3	9	13	12	12	350
O4	10	17	13	10	200
Requirements	450	310	200	190	

(*ii*)

Origin	Destination				Availability
	1	*2*	*3*	*4*	
O1	1	M	2	6	40
O2	5	8	1	M	20
O3	M	3	2	2	35
O4	10	7	3	1	20
Requirements	40	30	20	19	

Where **M** is a very large positive number.

(*iii*)

Available						Required
5	3	7	3	8	5	3
5	6	12	5	7	11	4
2	8	3	4	8	2	2
9	6	10	5	10	9	8
3	3	6	2	1	2	

17. The following table gives the cost of transporting material from supply points A, B, C, and D to demand points, E, F, G, H, and J.

	E	F	G	H	J
A	8	10	12	17	15
B	15	13	18	11	9
C	14	20	6	10	13
D	13	19	7	5	12

The present allocation is as follows: A to E 90; A to F 10; B to F 150; C to F 10; C to G 50; C to J 120; D to H 210; D to J 70.

(*i*) Check if this allocation is optimum. If not, find an optimum schedule. (6) (2) If in the previous problem the transportation cost from A to G is reduced to 10, what will be the new optimum schedule?

(ii) If in the previous problem the transportation cost from A to G is reduced to 10, what will the new optimum schedulebe?

18. XYZ Company collects milk from villagers every day and supplies them to three processing plants in Ghaziabad, Faridabad, and Sonepat. Milk is collected in vans from three village cooperatives starting at 4:00 A.M. in the morning. The time required (in hours) to transport the milk from the three supply points to the processing plants is given in the following table. To keep the calculations simple, the unit of measure is assumed as the number of vans.

Village Cooperatives	Ghaziabad	Faridabad	Sonepat	Supply (Vans)
A	5	7	8	15
B	6	5	5	20
C	5	8	7	30
Requirements (vans)	20	20	15	

Find the transportation plan.

19. A firm manufacturing a single product has plants I, II, and III. The three plants have produced 60, 35, and 40 units, respectively, during this month. The firm had made a commitment to sell 2 units to customer A, 45 units to customer B, 20 units to customer C, 18 units to customer D, and 30 units to customer E. Find the optimum transportation cost of shipping the manufactured product to five customers. The net per unit cost of transporting from the three plants to the five customers is given in the table.

Customer

		A	B	C	D	E
	I	4	1	3	4	4
Plant	**II**	2	3	2	2	3
	III	3	5	2	4	4

20. Solve the following profit-maximizing transportation problem using the best starting solution.

Origin	Destination				Availability
	1	*2*	*3*	*4*	
O1	4	5	8	6	100
O2	3	7	1	4	150
O3	9	5	2	2	150
Requirements	100	100	100	100	

21. What is a transshipment problem? How does it differ from the transportation problem? What are different types of transshipment problems?

22. Find the optimum solution to the following transshipment problems:

(*i*)

	S1	*S2*	*D1*	*D2*	Availability
S1	0	2	4	3	100
S2	2	0	5	6	150
D1	4	5	0	1	
D2	3	6	1	0	
Requirements			120	130	

(*ii*)

	D1	D2	D3	T1	T2	Availability
S1	∞	∞	∞	21	19	100
S2	∞	∞	∞	16	18	150
T1	15	14	12	0	∞	
T2	17	12	14	∞	0	
Requirements	50	100	100			

Write the LPP formulation of the transshipment problem given in Question 18.

CHAPTER 8

ASSIGNMENT PROBLEMS

8.1. Introduction

The transportation model discussed in the previous chapter tells how materials are to be shipped from different sources to various destinations. This model has a number of applications, and one such application is the assignment problem. An assignment problem is a special type of transportation problem in which a number of origins have to be assigned to the equal number of destinations such that the total cost (or time or distance) incurred in making this assignment is at a minimum; for example, "*n*" number of jobs have to be assigned to "*n*" number of people where the efficiency of doing different jobs by different people is different, and the assignment has to be done in such a way that the total time of completing all the jobs is at a minimum. Further, the assignment has to be made on a one-to-one basis; that is, one job should be assigned to one worker only and vice versa. Here the origins and destinations may be in the form of machines, people, plants, jobs, vehicles, and so forth. An assignment problem has its application in allocation and scheduling where, for example, planes or crews are assigned to commercial airline flights, vehicles or drivers are assigned to different routes, and so on.

8.2. Mathematical Formulation

Let there be *n* people who are to be assigned to *n* jobs, and c_{ij} denotes the measure of the performance of the ith person doing the jth job.

Define the decision variable x_{ij} ($i = 1, 2, 3, \ldots, n, j = 1, 2, 3, \ldots, n$), as follows—

$$x_{ij} = \begin{bmatrix} 1, & \text{if } i^{th} \text{ person gets the } j^{th} \text{ job} \\ 0, & \text{otherwise} \end{bmatrix}$$

So, x_{ij} is a binary variable which can take only two values, either 1 or 0.

The mathematical model for an assignment problem can be formulated as:

$$Min\ Z = \sum_{i=1}^{n} \sum_{j=1}^{n} c_{ij} x_{ij}$$

Subject to $\sum_{i=1}^{n} x_{ij} = 1;\ j = 1, 2, \ldots, n$

$$\sum_{j=1}^{n} x_{ij} = 1;\ i = 1, 2, \ldots, n \qquad \ldots(8.1)$$

$x_{ij} \geq 0 \ \forall \ i, j$ where $x_{ij} = 0$ or 1 and c_{ij} is the cost of assigning the jth job to the ith person.

The previous problem is a special type of linear programming problem where decision variables are restricted to take only integer binary values of either 0 or 1; for this reason (8.1) is sometimes known as a binary integer linear programming problem.

8.3. Assignment Problems as a Special Case of Transportation Problems

As mentioned previously, an assignment problem is one of the many applications of a transportation problem. Here the number of origins is made equal to the number of destinations. The functional constraints of (8.1) make sure that only one origin is assigned to each destination and only one destination is assigned to each origin. Hence, an assignment problem is formulated like an $n \times n$ transportation problem, where supply at each source is one and demand at each destination is also one. There are $(2 * n - 1)$ basic variables in this problem, out of which only n variables will have value 1, and others will be zero. The solution of the problem will be highly degenerate if it is solved by the transportation algorithm or the simplex algorithm. Hence, to solve an assignment problem, a special method called the **Hungarian method** is used to get the optimal assignments.

Tabular representation of an assignment problem as a special case of transportation problem is as follows:

Men		Jobs				Supply
		J_1	J_2		J_n	
	M_1	c_{11} x_{11}	c_{12} x_{12}	...	c_{1n} x_{1n}	1
	M_2	c_{21} x_{21}	c_{22} x_{22}	...	c_{2n} x_{2n}	1
	$\vdots$	$\vdots$	$\vdots$		$\vdots$	$\vdots$
	M_n	c_{n1} x_{n1}	c_{n2} x_{n2}	...	c_{nn} x_{nn}	1
Demand		1	1	...	1	

The assignment problem can also be represented in form of a network as follows:

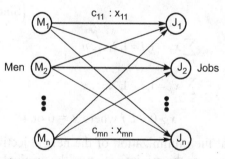

Fig. 8.1

Note: If the number of rows is not equal to the number of columns in the assignment problem, then that problem is called an unbalanced assignment problem and an appropriate number of dummy origins and dummy destinations are added to make it a balanced problem.

Theorem 8.1: In an assignment problem, if a constant is added or subtracted from all elements of the cost matrix $[c_{ij}]$, then the assignments that minimize the total cost on the original cost matrix will also minimize the new cost matrix; that is, the optimal assignments obtained for the changed matrix will also be optimal for the original matrix.

Proof: Let x_{ij} $(i = 1, 2, ..., n; j = 1, 2, ..., n)$ be the decision variables of the assignment problem, and then its mathematical formulation can be written as:

$$Min \ z = \sum_{i=1}^{n} \sum_{j=1}^{n} c_{ij} x_{ij}$$

Subject to
$$\sum_{i=1}^{n} x_{ij} = 1; \ j = 1, 2, ..., n$$

$$\sum_{j=1}^{n} x_{ij} = 1; \ i = 1, 2, ..., n$$

$$x_{ij} \geq 0 \ \forall \ i, j \text{ where } x_{ij} = 0 \text{ or } 1$$

Let $c_{ij}^* = c_{ij} \pm a$ (where "a" is any constant) be the new value of the cost associated with assigning the ith resource to the jth activity and z^* be the corresponding value of the objective function. Then,

$$Min \ z^* = \sum_{i=1}^{n} \sum_{j=1}^{n} \left(c_{ij} \pm a \right) x_{ij}$$

$$= \sum_{i=1}^{n} \sum_{j=1}^{n} c_{ij} x_{ij} \pm a \sum_{i=1}^{n} \sum_{j=1}^{n} x_{ij} = z \pm an$$

$$\left(\text{Since } \sum_{i=1}^{n} \sum_{j=1}^{n} x_{ij} = n \right)$$

Subject to
$$\sum_{i=1}^{n} x_{ij} = 1; \ i = 1, 2, ..., n$$

$$\sum_{j=1}^{n} x_{ij} = 1; \ i = 1, 2, ..., n$$

$$x_{ij} \geq 0 \ \forall \ i, j \text{ where } x_{ij} = 0 \text{ or } 1$$

This shows that the minimization of the new objective function z^* will have the same solution as the minimization of the original objective function, because *"an"* is a constant and is independent of the decision variables x_{ij}. But the new value of the objective function will be increased or decreased by *'an'* from the original value of z.

8.4. Hungarian Method

The steps for this computational procedure are as follows:

1. First, make sure that the assignment cost matrix is balanced, that is, the number of rows is equal to the number of columns. Add a sufficient number of dummy rows and columns if required to make it balanced.

2. Determine the minimum element from each row of the cost matrix and subtract that element from all elements of the corresponding row; this operation is called row reduction.

3. Determine the minimum element from each column of the changed matrix and subtract it from all elements of that corresponding column; this operation is called column reduction.

4. The reduced matrix will contain non-negative elements and at least one zero in each row and each column. Try to make assignments among zeros in the recently obtained matrix by the following procedure:

   (*a*) Starting with the first row of the cost matrix, find the row having a single zero, make an assignment in this zero by encircling it, and cross out all other zeros in the column containing this encircled zero. Continue searching another row having a single zero and repeat the procedure just described until all the rows of the matrix have been covered. Keep skipping the rows which have two or more zeros and, when only these types of rows are left in the matrix, repeat the same row operation on the columns.

   (*b*) Now, when we are left with the matrix having rows and columns (or only rows or only columns) with two or more zeros, then choose a row or column having a minimum number of zeros, arbitrarily encircle a zero in the chosen row or column, and cross out all zeros in the corresponding row and column containing the encircled zero. Repeat the whole procedure until all zeros are either encircled or crossed off.

   (*c*) If each row and each column of the obtained matrix has one and only one assignment with a zero, then we have the optimal assignment solution of the given problem; otherwise (if there are one or more rows or columns in the matrix having no assignment at all), use the following procedure:

      (*i*) Tick the row(s) which has no assignment (no encircled zero).

      (*ii*) In the ticked row(s), tick the column(s) having a crossed zero.

      (*iii*) In the ticked column(s), tick the row(s) having an encircled zero.

      (*iv*) The ticking process will continue as long as the previous conditions continue to be satisfied; otherwise, stop the process of ticking and draw lines covering all the ticked columns and un-ticked rows.

(v) Choose the minimum of the uncovered (by lines) elements, subtract it from all uncovered elements, add at the intersection of lines, and leave the other covered elements as is. Obtain the new modified matrix.

5. Repeat the steps (4) again. After obtaining the optimal solution, add the cost element in the original cost matrix at the position of the optimal assignment and determine the optimal cost.

Example 8.1. *Solve the following assignment problem using the* **Hungarian method.**

Coat (in $)

	Job1	Job2	Job3	Job4
Machine 1	7	10	5	4
Machine 2	9	5	8	11
Machine 3	15	9	4	7
Machine 4	10	6	12	4

Also give the linear programming formulation of the previous problem.

Solution: Since the number of rows equals the number of columns, the given assignment problem is a balanced one. We will explain the Hungarian algorithm using this example.

Step 1: Subtract row minima (row reduction).

We start with subtracting the row minimum from each row. For example, in the first row of the given matrix, the minimum element is 4; therefore, we subtract 4 from each element in the first row and the same will be done for the remaining rows. The resulting matrix is:

	Job1	Job2	Job3	Job4
Machine 1	3	6	1	0
Machine 2	4	0	3	6
Machine 3	11	5	0	3
Machine 4	6	2	8	0

Step 2: Subtract column minima (column reduction)

Similarly, the previous operation is repeated for the columns of the reduced matrix, and the following is the resulting matrix:

	Job1	Job2	Job3	Job4
Machine 1	0	6	1	0
Machine 2	1	0	3	6
Machine 3	8	5	0	3
Machine 4	3	2	8	0

Step 3: Making assignments.

We can now see that the previous matrix contains at least one zero in each row and each column. Therefore, we can start assigning the various jobs to the various machines. The first assignment is done to that row which contains a single zero, starting from the first row. In this case the second row contains a single zero, so we assign Job2 to Machine 2; this is done by encircling the zero and crossing off all the zeros (if any) in the column containing the encircled zero. Repeat the whole procedure until all zeros are either encircled or crossed off.

	Job1	Job2	Job3	Job4
Machine 1	⓪	6	1	⓪
Machine 2	1	⓪	3	6
Machine 3	8	5	⓪	3
Machine 4	3	2	8	⓪

Since each row and each column has exactly one assignment, the previous is the optimal solution of the given problem. Hence, Job 1 is assigned to Machine 1, Job 2 to Machine 2, Job 3 to Machine 3, and Job 4 to Machine 4. The total minimum cost of the assignment is $7 + 5 + 4 + 4 = \$20$.

Now the mathematical formulation of the previous assignment problem as an LPP is as follows:

Let x_{ij} ($i = 1, 2, 3, 4; j = 1, 2, 3, 4$) be the decision variable which denotes that the jth Job is assigned to the ith Machine. Then our problem is:

$$Min\ z = 7x_{11} + 10x_{12} + 5x_{13} + 4x_{14} + 9x_{21} + 5x_{22} + 8x_{23} + 11x_{24} + 15x_{31} + 9x_{32} + 4x_{33} + 7x_{33} + 10x_{41} + 6x_{42} + 12x_{43} + 4x_{44}$$

Subject to the constraints;

$$
\left.
\begin{aligned}
x_{11} + x_{12} + x_{13} + x_{14} &= 1 \\
x_{21} + x_{22} + x_{23} + x_{24} &= 1 \\
x_{31} + x_{32} + x_{33} + x_{34} &= 1 \\
x_{41} + x_{42} + x_{43} + x_{44} &= 1
\end{aligned}
\right\} \text{Row Constraints}
$$

$$
\left.
\begin{aligned}
x_{11} + x_{21} + x_{31} + x_{41} &= 1 \\
x_{12} + x_{22} + x_{32} + x_{42} &= 1 \\
x_{13} + x_{23} + x_{33} + x_{43} &= 1 \\
x_{14} + x_{24} + x_{34} + x_{44} &= 1
\end{aligned}
\right\} \text{Column Constraints}
$$

Whare $x_{ij} = 1$ or $0 \ \forall \ i, j$

Example 8.2. (Unbalanced Assignment problem) *Solve the following assignment problem using the Hungarian algorithm.*

Cost (in $)

	I	II	III
A	55	69	38
B	40	67	70
C	38	55	67
D	62	44	89

Solution: Since the number of rows is greater than the number of columns, the given problem is an unbalanced assignment problem. First we convert this unbalanced problem into a balanced one by introducing a new dummy column **IV**, with the cost cell entries as all zeros. The new matrix is given as follows:

	I	II	III	IV
A	55	69	38	0
B	40	67	70	0
C	38	55	67	0
D	62	44	89	0

Since the minimum of each row is zero, the rows will remain unchanged after the application of the row reduction procedure, and hence by performing the column reduction operations, we get:

$$
\begin{array}{c c c c c}
 & \textbf{I} & \textbf{II} & \textbf{III} & \textbf{IV} \\
\textbf{A} & 17 & 25 & 0 & 0 \\
\textbf{B} & 2 & 23 & 32 & 0 \\
\textbf{C} & 0 & 11 & 29 & 0 \\
\textbf{D} & 24 & 0 & 51 & 0
\end{array}
$$

Since the previous matrix contains at least one zero in each column and in each row, we can start assignments from the second row which contains only one zero.

$$
\begin{array}{c c c c c}
 & \textbf{I} & \textbf{II} & \textbf{III} & \textbf{IV} \\
\textbf{A} & 17 & 25 & ⓪ & ⓪ \\
\textbf{B} & 2 & 23 & 32 & ⓪ \\
\textbf{C} & ⓪ & 11 & 29 & ⓪ \\
\textbf{D} & 24 & ⓪ & 51 & ⓪
\end{array}
$$

Since each row and each column has exactly one assignment among zeros, the previous is the optimal solution of the given problem. Hence, **I** is assigned to **C**, **II** to **D**, **III** to A, and **B** will not get any job since B has been assigned job IV, which is a dummy job, and has been added just to solve the problem. The corresponding total minimum cost of assignment is 38 + 38 + 44 = $120.

Example 8.3. *A manager of a company has to assign three jobs, 1, 2, and 3 to three machines A, B, and C. The estimated job completion times in weeks are given in the following table.*

Time (in days)

	Job1	Job2	Job3
Machine A	27	12	20
Machine B	19	18	20
Machine C	24	14	17

The manager wishes to minimize the total number of weeks required to complete all three jobs. How should the allocation of machines to jobs be made?

Solution: Applying the row reduction operation in the given cost matrix, we have:

$$
\begin{array}{c}
\quad\quad\quad\quad\quad \textbf{Job1}\quad \textbf{Job2}\quad \textbf{Job3} \\
\begin{array}{l}
\textbf{Machine A} \\
\textbf{Machine B} \\
\textbf{Machine C}
\end{array}
\left(\begin{array}{ccc}
15 & 0 & 8 \\
1 & 0 & 2 \\
10 & 0 & 3
\end{array}\right)
\end{array}
$$

On applying the column reduction operation in the previous matrix, we have:

$$
\begin{array}{c}
\quad\quad\quad\quad\quad \textbf{Job1}\quad \textbf{Job2}\quad \textbf{Job3} \\
\begin{array}{l}
\textbf{Machine A} \\
\textbf{Machine B} \\
\textbf{Machine C}
\end{array}
\left(\begin{array}{ccc}
14 & 0 & 6 \\
0 & 0 & 0 \\
9 & 0 & 1
\end{array}\right)
\end{array}
$$

As the previous matrix contains at least one zero in each column and in each row, we can make assignments in rows and columns having single zeros and cross off all zeros in the respective column and row where the assignment has been made. So we get:

$$
\begin{array}{c}
\quad\quad\quad\quad\quad \textbf{Job1}\quad \textbf{Job2}\quad \textbf{Job3} \\
\begin{array}{l}
\textbf{Machine A} \\
\textbf{Machine B} \\
\textbf{Machine C}
\end{array}
\left(\begin{array}{ccc}
14 & ⓪ & 6 \\
⓪ & ⊗ & 2 \\
9 & ⊗ & 3
\end{array}\right)
\end{array}
$$

Now, since the number of assignments made is less than the number of rows/columns, the optimal assignment is yet to be reached. Now we shall draw a minimum number of lines to cover all the zeros so as to move toward the optimal solution (as given in the **Hungarian method**). The following are the steps for doing this:

(i) Tick row 3 since it has no assignment.

(ii) Tick column 2 since it has a crossed zero in the ticked row 3.

(iii) Tick row 1 since this has an assignment in the ticked column 2.

(iv) Since no other rows or columns can be further ticked, draw a line to all the ticked column(s) and un-ticked row(s).

$$
\begin{array}{c}
\quad\quad\quad \text{Job1} \quad \text{Job2} \quad \text{Job3} \\
\begin{array}{l}
\text{Machine A} \\
\text{Machine B} \\
\text{Machine C}
\end{array}
\left(
\begin{array}{ccc}
14 & ⓪ & 6 \\
⓪ & ⊗ & 2 \\
9 & ⊗ & 3
\end{array}
\right)
\begin{array}{l}
\checkmark \\
\\
\checkmark
\end{array}
\end{array}
$$

We now modify the matrix by subtracting 3 (which is the minimum among the elements uncovered by the lines) from all the uncovered elements and add the same to the elements appearing at the intersection of the lines. The modified matrix is shown as follows:

$$
\begin{array}{c}
\quad\quad\quad \text{Job1} \quad \text{Job2} \quad \text{Job3} \\
\begin{array}{l}
\text{Machine A} \\
\text{Machine B} \\
\text{Machine C}
\end{array}
\left(
\begin{array}{ccc}
11 & 0 & 3 \\
0 & 3 & 0 \\
6 & 0 & 0
\end{array}
\right)
\end{array}
$$

Now we repeat the previous procedure to find the new assignment:

$$
\begin{array}{c}
\quad\quad\quad \text{Job1} \quad \text{Job2} \quad \text{Job3} \\
\begin{array}{l}
\text{Machine A} \\
\text{Machine B} \\
\text{Machine C}
\end{array}
\left(
\begin{array}{ccc}
11 & ⓪ & 3 \\
⓪ & 3 & ⊗ \\
6 & ⊗ & ⓪
\end{array}
\right)
\end{array}
$$

Since the number of assignments is equal to the order of the given matrix, an optimum solution is attained:

$$\text{Machine A} \longrightarrow \text{Job 2}$$

$$\text{Machine B} \longrightarrow \text{Job 1}$$

$$\text{Machine C} \longrightarrow \text{Job 3}$$

Total minimum time for completing the three jobs by the three Machines is 48 days.

8.5. Special Cases in Assignment Methods

8.5.1. Maximization Problems

If the given problem is a maximization problem, then the **Hungarian method** cannot be used to find the optimal assignment to such problems, because the **Hungarian method** requires minimization of the objective function. To use the **Hungarian method** to find a solution to such problems, we must first convert the given profit matrix into an opportunity loss matrix, and this is done by subtracting all the elements of the matrix from the largest element of the matrix. Now the given maximization problem will convert to a minimization problem, and hence the **Hungarian method** can be used to find the optimal assignment. The corresponding maximum profit can be obtained by adding the elements present at the position of the optimal allocations in the original profit matrix.

Example 8.4. *A company selling five different products has five different buyers. The ith buyer is willing to pay price p_{ij} (or less) for the jth product. The values of p_{ij} are given in the following matrix. The company wishes to know which product to offer to each buyer so as to maximize total profit.*

$$
\text{Buyer}
\begin{array}{c}
\\
\\
\\
\\
\\
\end{array}
\overset{\text{Products}}{\begin{pmatrix}
12 & 15 & 24 & 10 & 29 \\
20 & 15 & 30 & 19 & 24 \\
33 & 21 & 22 & 18 & 27 \\
15 & 19 & 20 & 24 & 16 \\
20 & 27 & 21 & 26 & 25
\end{pmatrix}}
$$

Solution. The given matrix contains per unit profit; it is not a cost matrix as was considered in earlier problems. Here the objective function is to be maximized. Therefore, in order to use the **Hungarian method** and to find an optimal assignment to the given problem, we need to convert the given matrix into an opportunity loss matrix by subtracting all the elements from the largest element of the matrix. In the given problem, the maximum element is **33**. The following is the resulting matrix.

$$
\begin{pmatrix}
21 & 18 & 9 & 23 & 4 \\
13 & 18 & 3 & 14 & 9 \\
0 & 12 & 11 & 15 & 6 \\
18 & 14 & 13 & 9 & 17 \\
13 & 6 & 12 & 7 & 8
\end{pmatrix}
$$

On applying the row and column reduction method, we have the following matrix:

$$\begin{pmatrix} 17 & 14 & 5 & 19 & 0 \\ 10 & 15 & 0 & 11 & 6 \\ 0 & 12 & 11 & 15 & 6 \\ 9 & 5 & 4 & 0 & 8 \\ 7 & 0 & 6 & 1 & 2 \end{pmatrix}$$

Since each row and each column has at least one zero, we can now start putting in the assignments:

$$\begin{pmatrix} 17 & 14 & 5 & 19 & ⓪ \\ 10 & 15 & ⓪ & 11 & 6 \\ ⓪ & 12 & 11 & 15 & 6 \\ 9 & 5 & 4 & ⓪ & 8 \\ 7 & ⓪ & 6 & 1 & 2 \end{pmatrix}$$

Since the total number of assignments in the matrix is equal to the number of rows/columns, the present assignments are the optimal assignments. So, buyer 1 should be given product 5, buyer 2 should be given product 3, buyer 3 should be given product 4, and finally buyer 5 should be given product 2. The corresponding maximum profit is $29 + 30 + 33 + 24 + 27 = 143$.

8.5.2. Restricted Entry

If the assignment at any position of the matrix is prohibited (it is restricted or not allowed due to some reasons), for example, the ith origin cannot be assigned to the jth destination due to some circumstances, then assign infinity (∞) at that location and follow the Hungarian method to get the optimal assignment schedule and the optimal value of the measure of effectiveness.

8.6. Solution of Assignment Problems Using Excel Solver

An assignment problem can also be solved using MS-Excel in the same manner as the transportation problem was solved. The only difference is in this case, the availability or supply at each source and demand of each destination will be exactly equal to unity. Consider Example 8.1.

Coat (in $)

	Job1	Job2	Job3	Job4
Machine 1	7	10	5	4
Machine 2	9	5	8	11
Machine 3	15	9	4	7
Machine 4	10	6	12	4

1. Create a solution space on the same Excel sheet.

	Cost (in $)			
	Job 1	**Job 2**	**Job 3**	**Job 4**
Machine1	7	10	5	4
Machine2	9	5	8	11
Machine3	15	9	4	7
Machine4	10	6	12	4

	Job 1	**Job 2**	**Job 3**	**Job 4**	**Supply**
Machine1					1
Machine2		Space for solution			1
Machine3					1
Machine4					1
Demand	1	1	1	1	

2. Insert a blank column and row alongside the space provided for solution (x_{ij}) to compute the row and column sum, as shown in the following figure:

	Cost (in $)			
	Job 1	**Job 2**	**Job 3**	**Job 4**
Machine1	7	10	5	4
Machine2	9	5	8	11
Machine3	15	9	4	7
Machine4	10	6	12	4

	Job 1	**Job 2**	**Job 3**	**Job 4**	**Row sum**	**Supply**
Machine1					0	1
Machine2					0	1
Machine3					0	1
Machine4					0	1
Column sum	=SUM(C10:C14)		0	0		
Demand	SUM(number1, [number2], ...)			1		

3. Create a space for computing the value of the total minimum cost of the assignment. The **SUMPRODUCT** command will be used to multiply the given unit cost and solution space.

	Cost (in $)			
	Job 1	Job 2	Job 3	Job 4
Machine1	7	10	5	4
Machine2	9	5	8	11
Machine3	15	9	4	7
Machine4	10	6	12	4

	Job 1	Job 2	Job 3	Job 4	Row sum	Supply
Machine1					0	1
Machine2					0	1
Machine3					0	1
Machine4					0	1
Column sum	0	0	0	0		
Demand	1	1	1	1		

Total minimum Assignment cost =SUMPRODUCT(C4:F7,C10:F13)

SUMPRODUCT(array1, [array2], [array3], [array4], ...)

4. Now go to Solver. On solving the problem in the same manner as we had done in transportation, we have the following as the optimal solution and its corresponding minimum assignment cost:

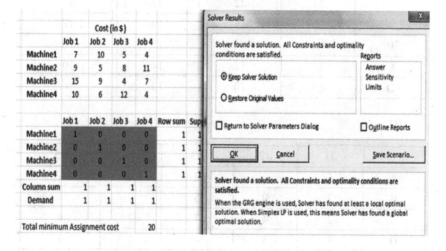

Exercises

1. Explain assignment problems as a particular case of a transportation problem.

2. What is an assignment problem? Give some real-life example.

3. Give the LPP formulation of an assignment problem.

4. Explain the limitation of using the simplex method for solving an assignment problem.

5. Explain the steps involved in the Hungarian algorithm for solving an assignment problem.

6. If a fixed constant is added to every element of the cost matrix, then show that the optimal assignments obtained for the changed matrix will also be optimal for the original matrix.

7. Explain the steps for solving a profit maximizing assignment problem.

8. Solve the following cost minimizing assignment problem:

(*i*)

	J1	J2	J3	J4
A	20	15	10	17
B	9	14	11	18
C	13	12	14	10
D	15	10	13	9

(*ii*)

	J1	J2	J3	J4	J5
A	6	–	2	4	1
B	3	9	5	2	–
C	2	4	7	3	6
D	4	3	–	4	1
E	7	5	8	7	5

(*iii*)

	J1	J2	J3	J4	J5
A	200	150	100	140	120
B	109	156	210	180	166
C	140	143	164	180	175
D	140	150	165	173	188

9. Consider the problem of assigning five jobs to five persons. The assignment costs are given as follows:

	J1	J2	J3	J4	J5
A	8	4	2	6	1
B	0	9	5	5	4
C	3	8	9	2	6
D	4	3	1	0	3
E	9	5	8	9	5

Determine the optimum assignment schedule.

10. Find an optimal solution to an assignment problem with the following cost matrix:

	J1	J2	J3	J4
M1	10	9	7	8
M2	5	8	7	7
M3	5	4	6	5
M4	2	3	4	5

11. A company has four sales representatives who are to be assigned to four different sales territories. The monthly sales increases estimated for each

sales representative for different sales territories (in lakhs of Rupees) are shown in the following table:

Sales Representatives	Sales Territories			
	I	II	III	IV
A	200	150	170	220
B	160	120	150	140
C	190	195	190	200
D	180	175	160	190

Suggest optimal assignments and the total maximum sales increase per month. If for a certain reason, sales representative B cannot be assigned to sales territory III, will the optimal assignment schedule be different? If so, find that schedule and effect on total cost.

12. An automobile dealer wishes to send four repairmen to four different jobs. The repairmen have somewhat different kinds of skills, and they exhibit different levels of efficiency from one job to another. The dealer has estimated the number of man-hours that would be required for each job-man combination. This is given in the matrix form in the following table. Find the optimum assignment that will result in the minimum man-hours needed.

Job

	I	II	III	IV
A	5	3	2	8
B	7	9	2	6
C	6	4	5	7
D	5	7	7	8

Man

CHAPTER 9

GAME THEORY

9.1. Introduction

Game theory is a subject which deals with competitive situations and explains the art of decision making, where each competitor called player makes a choice that potentially affects the interests of the other competitors or players and promotes his own interest. These players may be individuals, groups, firms, or any combination of these. Game theory is different from decision analysis in the sense that in the latter case a game is played by the decision maker whose opponent is passive in nature and makes choices or decisions without having any interest in the other's loss.

This subject has its origin in the early half of the nineteenth century with the publication by Antoine Augustin Cournot in 1838 in *Researches into the Mathematical Principles of the Theory of Wealth,* in which he had attempted to explain the underlying rules governing the behavior of duopolists. Almost a century later, in the year 1921 the mathematician Emile Borel for the first time suggested a formal theory of games. However, it was only in 1944 that the modern principles of game theory were formulated by John von Neumann and Oskar Morgenstern with their publication, *Theory of Games and Economic Behavior.* Much of the basic terminology that is in use today was published in this book.

Any competitive situation will be called a "game" if it has the following characteristics:

(i) There are finite numbers of competitors called players.

(ii) There are finite numbers of possible courses of action called strategies, available to each player. Here the strategy may be a simple action or a

predetermined rule which tells the player how to respond in different situations at each stage of the game.

(iii) There is a defined rule which governs the conduct of the players.

(iv) The outcome of all the possible combinations of the courses of action taken by the various players determines the payoff to each player, which denotes win, lose, or draw.

When the game is played, each player knows the strategies available to him and the strategies available to his opponents as well as the payoff to him as a result of the combination of different strategies adopted by different players, but none of them exactly know the choice made by the other, and they simultaneously make their own best choice. The objective here is to develop the criterion to enable the players to select their best possible strategy/strategies.

9.2. Zero-Sum Games

Any competitive game can be classified as a "zero-sum" or "non-zero-sum" game on the basis of its outcome. If the algebraic sum of the gains and losses of all the players involved in the game is zero, then the game is said to be a "zero-sum" game; otherwise, it's a "non-zero-sum" game. In a zero-sum game there is no addition or reduction in the total wealth of all the players; instead, the same total wealth is redistributed among the competing players.

If the given game problem consists of only two players, then it's called a two-person game. In a two-person game if the gain of one player is equal to the loss of the other player, then it's called a two-person zero-sum game, and it's also assumed that each player knows the outcome for all possible combinations of strategies that he and his competitor may have adopted during the course of the game. The resulting outcomes in a two-person zero-sum game can be displayed in the form of a matrix called a payoff matrix. $(a_{ij})_{m \times n}$; here a_{ij} represents the payoff or utility which may be in terms of monetary benefit, and also when one player selects the ith strategy while j^{th} is chosen by other, and the possible number of strategies available to both the players are m and n. In other words, if a game is played by two players A and B, then this is the payoff matrix for player A, where a_{ij} is the amount paid to A by B if A plays his i^{th} strategy and B plays his j^{th} strategy, and hence the elements of the payoff matrix denote the resulting gain (loss) to a particular player. In a two-person zero-sum game, if a player A has a total of m strategies and the other player B

has a total n strategies available to him, then the resulting payoff of the game can be put in the form of an $m \times n$ payoff matrix (a_{ij}), where each element of the matrix represents the various possible values of the game when the two players select their different strategies. In a zero-sum two-person game, the payoff matrix for player B will be the negative of the payoff matrix of player A; that is, the gain of one will be the loss of the other, such that the sum is zero. To understand this in a better way, let us consider the following example involving two different players A and B. Let player A have a total of 3 different strategies denoted by I, II, and III, and player B has a total of 2 different strategies denoted by 1 and 2, and let the resulting payoff to player A be as given in the following 3×2 payoff matrix:

$$
\begin{array}{c}
B \\
\begin{array}{cc} 1 & 2 \end{array} \\
A\begin{array}{c} \text{I} \\ \text{II} \\ \text{III} \end{array}\begin{pmatrix} 1 & 4 \\ 4 & -1 \\ 0 & 1 \end{pmatrix}
\end{array}
$$

As per the previous payoff, player A is going to win a sum of \$4 if he selects his second strategy at the same time player B select his first strategy. Similarly, he will lose \$1 if he sticks to strategy II and player B selects strategy 2. The same interpretation applies to the remaining elements of the payoff matrix. Since it's a zero-sum game, the payoff matrix to player B can be written as:

$$
\begin{array}{c}
B \\
\begin{array}{cc} 1 & 2 \end{array} \\
A\begin{array}{c} \text{I} \\ \text{II} \\ \text{III} \end{array}\begin{pmatrix} -1 & -4 \\ -4 & 1 \\ 0 & -1 \end{pmatrix}
\end{array}
$$

The previous payoff is obtained by simply putting a negative ahead of every element of the payoff to player A. When we sum up the two payoffs, the resulting matrix is a null matrix.

In a two-person zero-sum game, the two players will be called a row player and a column player. The row designations are the courses of action available to the row player and the column designations are the courses of action available to the column player. The row player will always try to maximize his yield and for this reason, a row player is also sometimes known as a maximizing player, whereas the column player is a minimizing player and will try to minimize his loss. The payoff given in any game problem will always be considered for the maximizing player unless otherwise stated.

Note. If (a_{ij}) is the payoff to one player in a two-person zero-sum game, then $(-a_{ij})$ will be the payoff to the other player. The given payoff will always be assumed for player A (maximizing player) unless otherwise stated.

9.3. Maximin and Minimax Principle

In the maximin and minimax principle, the players adopt a conservative approach. The maximin (minimax) principle is a principle adopted by the maximizing (minimizing) player. Here the player takes a pessimistic approach and plays it safe by selecting the best out of the worst possible outcomes, that is, the player will always play that strategy which corresponds to the maximum (minimum) of the minimum (maximum) gains (losses) from the list of options available to him. Since both players are rational, if one player chooses one particular strategy, then the other player will act rationally and chooses the strategy in such a way that it will prevent his maximum loss or prevent his opponent to make maximum profit. Both the players are averse to taking the risk of getting larger losses and will play the moves which will guarantee them the best payoff and hence will result in a break-even position. This principle helps both the players in selecting the strategies which will be best for both of them in a way that even if one knows his/her opponent's selection of strategy, he won't be able to improve his loss or gain any more. To understand it better, consider a two-person zero-sum game with the following payoff matrix:

$$
\begin{array}{c}
\textbf{Player } B \\
\begin{array}{cccc}
I & II & III & IV
\end{array} \\
\text{Player } A \begin{array}{c} I \\ II \\ III \end{array}
\begin{pmatrix}
2 & 3 & 5 & 1 \\
6 & 4 & 6 & 5 \\
4 & 2 & 7 & 9
\end{pmatrix}
\end{array}
$$

Here player A is a maximizing player and has a total of three different strategies available to him, whereas B is the minimizing player and has a total of four different strategies available to him. When player A plays his strategies, we have the following:

(i) Worst outcome when A plays I = min (1st row) = min (2, 3, 5, 1) = 1.

(ii) Worst outcome when A plays II = min (2nd row) = min (6, 4, 6, 5) = 4.

(iii) Worst outcome when A plays III = min (3rd row) = min (4, 2, 7, 9) = 2.

Now the best of these worst outcomes is max (1, 4, 2) = 4 . Hence, the maximin for the maximizing player A is 4. Similarly, the worst possible outcomes of the minimizing player B are

(i) Worst outcome when B plays I = max (1st column) = max (2, 6, 4) = 6.

(ii) Worst outcome when B plays II = max (2nd column) = max (3, 4, 2) = 4.

(iii) Worst outcome when B plays III = max (3rd column) = max (5, 6, 7) = 7.

(iv) Worst outcome when B plays IV = max (4th column) = max (1, 5, 9) = 9.

And the best of these worst outcomes for player B is min (6, 4, 7, 9) = 4. Hence, the minimax for the minimizing player B is also 4.

Saddle Point. A saddle point of payoff matrix is that point of the matrix where the maximin value equals the minimax value of the matrix. This point of the matrix is also known as the equilibrium point of the matrix. The value of the payoff at the saddle point is called the value of the game, and the corresponding strategies are the optimal strategies of the two players. Such a game is said to be a deterministic game.

In the previous example, $\min_j \max_i (a_{ij}) = \max_i \min_j (a_{ij}) = 4$. Hence, 4 is the saddle point of the game and it's the value of the game. The optimal strategy for player A is II and that of B is also II. Hence, if the saddle point exists in the game, then both players should adopt the strategies determined by the saddle point, since it gives the stable solution where none of them can take advantage of his/her opponent to improve his payoff.

It is not necessary that the saddle point of a game always exists. For those games without a saddle point, the minimax and maximin value of the payoff matrix will provide an upper and lower bound to the expected value of the game, obtained using a suitable method; that is, if v is the expected value of the game, then

$$\text{maximin value} \leq v \leq \text{minimax value}$$

Pure Strategy: A strategy is said to be a pure strategy if a player selects the same strategy each time regardless of the other player playing any of his strategies. In such situations each player knows in advance what the other player is going to do. In a game with a saddle point, each player plays their best strategy with a probability of 1.

Optimum Strategy: This is the best possible strategy (strategies) available to both the players.

Value of the game. It is the expected amount of gain (loss) to the maximizing (minimizing) player when both players play optimally. If the value of the game is zero, then the game is said to be a **fair game**.

Theorem 9.1. For any two-person zero-sum game with $(a_{ij})_{m \times n}$ as its payoff, the maximin value $\underline{v}$ of the matrix will always be less or equal to the minimax value $\bar{v}$ of the matrix; that is, $\max_i \min_j (a_{ij}) \le m_i n_j \max_i (a_{ij})$.

Proof: We can easily see that,

$$\min_j (a_{ij}) \le a_{ij}, \ \forall_i \qquad\qquad \text{...(9.1)}$$

And $\quad \max_i (a_{ij}) \ge a_{ij}, \ \forall j \qquad\qquad \text{...(9.2)}$

Letting $\quad \min_j (a_{ij}) = a_{is}$ and $\max_j (a_{ij}) = a_{rj}$, we have

$$a_{is} \le a_{ij} \le a_{rj} \ i = 1, 2, 3. \dots, m \text{ and } j = 1, 2, 3. \dots, n \qquad \text{...(9.3)}$$

Now we can easily have,

$$\max_i a_{is} \le a_{ij} \le \min_j a_{rj}, \ i = 1, 2, 3. \dots, m \text{ and } j = 1, 2, 3. \dots, n \qquad \text{...(9.4)}$$

Hence $\max_i \min_j (a_{ij}) \le a_{ij} \le \min_j \max_i (a_{ij})$

$\Rightarrow \qquad \max_i \min_j (a_{ij}) \le \min_j \max_i (a_{ij}) \ \Rightarrow \ \underline{v} \le \bar{v}$

Remark

1. If the game possesses a saddle point, then $\underline{v} = v = \bar{v}$, and as mentioned previously, such games are said to be deterministic games, and each player will select a single strategy as their respective optimal strategy.

2. If $\underline{v} = v = \bar{v} = 0$, then the game is a fair game.

9.4. Game with a Saddle Point

The following are the steps involved for finding a saddle point of a game:

(*a*) Select the minimum element in each row of the payoff matrix and put them under the "row minima" column. Each of these minima is put inside $\bigcirc$, as shown as follows, in the matrix.

(*b*) Select the maximum element in each column of the payoff matrix and put them under the "column maximum" heading. Each of these maxima is put inside ☐ as shown as follows in the matrix.

(c) Now the element(s) with both ◯ and ☐ in the payoff matrix will represent the saddle point of the game, and hence the value of the game and the corresponding strategy of the two players are their respective optimal strategies.

Note: If the saddle point of the game exists, then it will represent the value of the game and the corresponding strategies are the optimal strategies. Theorem 9.1 will establish the relation between the minimax and maximin value for those games which don't possess a saddle point.

Example 9.1. *Check which of the following two-person zero-sum game problems represents a deterministic game. Find the optimum strategies for each player, determine if it is a deterministic game, and also mention if the game is fair.*

$$
(i) \quad \begin{array}{c} & & B \\ & & \text{I} \quad \text{II} \quad \text{III} \\ A \quad \begin{array}{c} \text{I} \\ \text{II} \\ \text{III} \end{array} & \begin{pmatrix} 2 & 3 & 5 \\ 6 & 4 & 6 \\ 4 & 2 & 7 \end{pmatrix} \end{array}
$$

$$
(ii) \quad \begin{array}{c} & & B \\ & & \text{I} \quad \text{II} \\ A \quad \begin{array}{c} \text{I} \\ \text{II} \\ \text{III} \end{array} & \begin{pmatrix} -1 & 1 \\ -2 & 2 \\ 0 & 1 \end{pmatrix} \end{array}
$$

$$
(iii) \quad \begin{array}{c} & & B \\ & & \text{I} \quad \text{II} \\ A \quad \begin{array}{c} \text{I} \\ \text{II} \\ \text{III} \end{array} & \begin{pmatrix} 1 & 2 \\ 2 & -1 \\ 0 & 1 \end{pmatrix} \end{array}
$$

(i) **Solution:** The minimum of each row is given under the "row minima" heading, and the corresponding element is encircled ◯ in the payoff matrix. Similarly, the maximum of each column is written under the heading "column maxima" and is shown inside ☐ in the matrix. The maxima of

these "row minima" will give the maximin value for player A, whereas the minimum of the "column maxima" will provide the minimax value for player B.

		B			
		I	II	III	Row minima
A	I	(2)	3	5	2
	II	6	(4)	6	4
	III	4	(2)	6	2
Column maxima		6	4	6	

Here for the given game, maximin = 4 = minimax, and this can be clearly seen as the crossing point of the ○ and □. Hence, the game has a saddle point, and the value of the game is 4. The best strategy for A and B are their respective second (II) strategies. This game is a deterministic game, and since its value is not equal to zero, it is an unfair game.

(ii) **Solution:** Here, we have maximin = 0 = minimax as shown clearly in the figure of payoff table below, and hence the game has a saddle point, and the value of the game is 0. Also, it is a fair and deterministic game. The best strategy for A is III and for B is I.

		B		
		I	II	Row minima
A	I	(−1)	1	−1
	II	(−2)	2	−2
	III	(0)	1	0
Column maxima		0	2	

(iii) **Solution:** For the given game, minimax = min (2, 2) = 2 and maximin = max (1, −1, 0) = 1. Since the two are not equal, the game doesn't possess a saddle point and hence it is not a deterministic game. This has to be solved using a different method, to be discussed in the following sections.

		B		
		I	II	Row minima
A	I	②(2)	⬜2	1
	II	⬜2	⬭(−1)	−1
	III	⬭(0)	1	0
Column maxima		2	2	

9.5. Game without a Saddle Point

For every two-person zero-sum game problem, it is not necessary that the maximin value will always equal the minimax value of the game and hence, in such situations, the game problem no longer remains a deterministic game, and the maximin-minimax principle for solving the game fails. In this case if both the players try to select the strategy as suggested by maximin and minimax criterion and at the same time predict the other player's strategy, though which are not the same, it will result in an unstable situation since both are rational players and will keep on changing their strategies in order to prevent his/her opponent to make a larger gain and to avoid incurring his own loss. Both the players will try to improve their own payoff. In this situation, the strategy should not be selected by some fixed criterion so that none of them should be able to predict the other's strategy, and both the players should make their choices randomly without taking advantage of the opponent's choice.

To solve such a game problem, the concept of a chance move is introduced. Here the two players will be selecting their different strategies according to a probability distribution. The objective of each player will be to select a set of strategies in a random manner so as to optimize his average payoff. The strategies so determined are called **mixed strategies** because each strategy is being selected according to a probability law assigned to it.

Mixed Strategy: If a game does not possess a saddle point, game theory suggests that each player assign a probability distribution to the set of strategies available to him. Then the players select at least two or more strategies with certain probabilities (weights). These probabilities (weights) represent how frequently each move is to be played by the two players. A mixed strategy is used when the player is indifferent among several pure strategies, and thus this keeps the opponent guessing the possible moves he will make.

Let $(a_{ij})_{m \times n}$ denote the payoff matrix of a two-person (A and B) zero-sum game where m & n are the maximum number of available strategies to players A and B, respectively. Further let x_i ($i = 1, 2, 3, \ldots, m$) and y_j ($j = 1, 2, 3, \ldots, n$) be the probabilities of player A playing the ith strategy, and player B playing the jth strategy, respectively, and then the expected (average) payoff to player A is given by:

$$E(x,y) = \sum_{i=1}^{m} \sum_{j=1}^{n} x_i y_j a_{ij} \qquad \ldots(9.5)$$

Here both the players will select mixed strategies according to random observations drawn from the probability distribution of their own strategies, such that Player A tries to maximize his minimum expected payoff, and player B minimizes his maximum expected loss. The optimal mixed strategies for both the players will provide a stable solution where:

Maximum of Minimum expected payoff for player A = Minimum of Maximum expected loss of player B = value of the game

None of them will be able to improve their positions by deviating from these optimal strategies.

Theorem 9.2. For any two-person (A and B) zero-sum game without a saddle point, with $\begin{pmatrix} a_{11} & a_{12} \\ a_{21} & a_{22} \end{pmatrix}$ as the payoff matrix to player A.

The optimum mixed strategies $S_A = \begin{bmatrix} A_1 & A_2 \\ x_1 & x_2 \end{bmatrix}$ and $S_B = \begin{bmatrix} B_1 & B_2 \\ y_1 & y_2 \end{bmatrix}$ are

given by $x_1 = \dfrac{a_{22} - a_{21}}{a_{11} + a_{22} - (a_{12} + a_{21})}$, $x_2 = (1 - x_1) = \dfrac{a_{11} - a_{12}}{a_{11} + a_{22} - (a_{12} + a_{21})}$

$y_1 = \dfrac{a_{22} - a_{12}}{a_{11} + a_{22} - (a_{12} + a_{21})}$ and $y_2 = (1 - y_1) = \dfrac{a_{11} - a_{21}}{a_{11} + a_{22} - (a_{12} + a_{21})}$.

The expected gain for player A is given by $v = \dfrac{a_{11}a_{22} - a_{12}a_{21}}{a_{11} + a_{22} - (a_{12} + a_{21})}$.

Proof: Given the mixed strategy for player A as $S_A = \begin{bmatrix} A_1 & A_2 \\ x_1 & x_2 \end{bmatrix}$, with $x_1 + x_2 = 1$.

If player B moves his strategy B_1, then the expected gain to player A will be:

$$E_1(x) = x_1 a_{11} + x_2 a_{21} = +x_1 a_{11} + (1-x_1) a_{21} = x_1(a_{11} - a_{21}) + a_{21} \quad ...(9.6)$$

and if B moves B_2, the expected gain to player A will be:

$$E_2(x) = x_1 a_{12} + x_2 a_{22} = x_1 a_{12} + (1-x_1) a_{22} = x_1(a_{12} - a_{22}) + a_{22} \quad ...(9.7)$$

Similarly, if B has a mixed strategy as $S_B = \begin{bmatrix} B_1 & B_2 \\ y_1 & y_2 \end{bmatrix}$ with $y_1 + y_2 = 1$, then the expected losses for Player B, when A plays his strategies A_1 and A_2, is:

$$E_1(y) = y_1 a_{11} + y_2 a_{12} \qquad ...(9.8)$$

$$E_2(y) = y_1 a_{21} + y_2 a_{22}, \text{respectively.} \qquad ...(9.9)$$

From (9.4), the total expected payoff to player A when B moves his different strategies with probabilities y_1 and y_2 is given by:

$$E(x, y) = y_1(a_{11}x_1 + a_{21}x_2) + y_2(a_{12}x_1 + a_{22}x_2)$$

$$= y_1(a_{11}x_1 + a_{21}(1-x_1)) + (1-y_1)(a_{12}x_1 + a_{22}(1-x_1)) \quad ...(9.10)$$

Let v denotes the expected payoff to player A (value of the game) when both the players play their optimum strategies. Now, in order that player A is unaffected with whatever choice of strategies player B makes, we must have:

$$x_1(a_{11} - a_{21}) + a_{21} = x_1(a_{12} - a_{22}) + a_{22}$$

$$\Rightarrow \qquad x_1 = \frac{a_{22} - a_{21}}{a_{11} + a_{22} - (a_{12} + a_{21})}$$

And $\qquad x_2 = (1-x)$

$$= 1 - \frac{a_{22} - a_{21}}{a_{11} + a_{22} - (a_{12} + a_{21})}$$

$$= \frac{a_{11} - a_{12}}{a_{11} + a_{22} - (a_{12} + a_{21})}$$

Similarly, from (9.7) and (9.8), we have:

$$y_1 = \frac{a_{22} - a_{12}}{a_{11} + a_{22} - (a_{12} + a_{21})}$$

and
$$y_2 = \frac{a_{11} - a_{21}}{a_{11} + a_{22} - (a_{12} + a_{21})}$$

Now putting the values of x_1 and y_1 in (9.9), we get:

$$E(x, y) = v = \frac{a_{11}a_{22} - a_{12}a_{21}}{a_{11} + a_{22} - (a_{12} + a_{21})}.$$

Example 9.2. *Solve the following game problems.*

(i)

$$\begin{array}{c} & B \\ & \begin{array}{cc} I & II \end{array} \\ A \begin{array}{c} I \\ II \end{array} \begin{pmatrix} 4 & 2 \\ 3 & 5 \end{pmatrix} \end{array}$$

(ii)

$$\begin{array}{c} & B \\ & \begin{array}{cc} B_1 & B_2 \end{array} \\ A \begin{array}{c} A_1 \\ A_2 \end{array} \begin{pmatrix} 1 & 10 \\ 6 & 5 \end{pmatrix} \end{array}$$

(i) **Solution:** Clearly the game does not possess a saddle point.

Let $S_A = \begin{bmatrix} I & II \\ x_1 & x_2 \end{bmatrix}$

and $S_B = \begin{bmatrix} I & II \\ y_1 & y_2 \end{bmatrix}$ be the mixed strategies of player A and B, respectively.

Then by theorem 9.2, we have

$$x_1 = \frac{a_{22} - a_{21}}{a_{11} + a_{22} - (a_{12} + a_{21})}$$

$$= \frac{5 - 3}{4 + 5 - (2 + 3)} = \frac{2}{4} = 0.5$$

and $x_2 = 1 - x_1 = 1 - 0.5 = 0.5.$

Similarly, $y_1 = \frac{a_{22} - a_{12}}{a_{11} + a_{22} - (a_{12} + a_{21})} = \frac{3}{4} = 0.75$

and $y_2 = 1 - y_1 = 0.25$ and the value of the game is,

$$v = \frac{a_{11}a_{22} - a_{21}a_{12}}{a_{11} + a_{22} - (a_{12} + a_{21})} = \frac{14}{4} = 3.5$$

Hence, the solution of the game is:

(*a*) The optimal mixed strategy of player *A* is (0.5   0.5)

(*b*) The optimal mixed strategy of player *B* is (0.75 0.25)

(*c*) The expected value of the game is $v = 3.5$

(*ii*) **Solution.** Maximin of the game is 5 and Minimax is 6, and hence there is no saddle point to the game.

Let $S_A = \begin{bmatrix} A_1 & A_2 \\ x_1 & x_2 \end{bmatrix}$ and $S_B = \begin{bmatrix} B_1 & B_2 \\ y_1 & y_2 \end{bmatrix}$ be the mixed strategies of player *A* and *B*, respectively.

Then by theorem 9.2, we have

$$x_1 = \frac{a_{22} - a_{21}}{a_{11} + a_{22} - (a_{12} + a_{21})} = \frac{-1}{-10} = 0.1$$

and $x_2 = 1 - x_1 = 1 - 0.1 = 0.9.$

Similarly, $y_1 = \frac{a_{22} - a_{12}}{a_{11} + a_{22} - (a_{12} + a_{21})} = \frac{-5}{-10} = 0.5$

and $y_2 = 1 - y_1 = 0.5$ and the value of the game is,

$$v = \frac{a_{11}a_{22} - a_{21}a_{12}}{a_{11} + a_{22} - (a_{12} + a_{21})} = \frac{-55}{-10} = 5.5$$

Hence, the solution of the game is:

(*d*) The optimal mixed strategy of player A is (0.1 0.9)

(*e*) The optimal mixed strategy of player B is (0.5 0.5)

(*f*) The expected value of the game is $v = 5.5$

9.6. Graphical Solution of 2 × *n* and *m* × 2 Games

The formulas given in theorem 9.2 are applicable only for those game problems having no saddle point and whose payoff matrix is of dimension 2 × 2. However, this procedure can be extended to a square payoff matrix of any dimension, but the difficulty arises when the payoff matrix is not a square.

In this section, we will discuss how to solve a game problem with a payoff matrix of the order 2 × *n* and *m* × 2. The technique is to reduce the dimension of the payoff matrix from $2 \times n$ and $m \times 2$ to 2×2 by graphically locating the optimal strategies and then applying theorem 9.2 to finally find the values. Consider an $m \times 2$ game whose payoff matrix is given by:

$$
\begin{array}{c}
 & B \\
 & \begin{array}{cc} B_1 & B_2 \end{array} \\
A \begin{array}{c} A_1 \\ A_2 \\ \vdots \\ A_m \end{array} & \begin{pmatrix} a_{11} & a_{12} \\ a_{21} & a_{22} \\ \vdots & \vdots \\ a_{m1} & a_{m2} \end{pmatrix}
\end{array}
$$

To start with, let us assume that the game does not possess a saddle point. Also let y_1 and y_2 denote the probabilities of player B playing his strategies B_1 and B_2, respectively, where $y_1 + y_2 = 1$. Now, the expected payoff to player B, when A plays his different strategies, is given by:

When A plays	Expected payoff to player B
A_1	$E_1(y) = y_1 a_{11} + y_2 a_{12} = y_1 a_{11} + (1 - y_1) a_{12}$
A_2	$E_2(y) = y_1 a_{21} + y_2 a_{22} = y_1 a_{21} + (1 - y_1) a_{22}$
$\vdots$	$\vdots$
A_m	$E_m(y) = y_1 a_{m1} + y_2 a_{m2} = y_1 a_{m1} + (1 - y_1) a_{m2}$

According to the minimax criterion, player B will determine the values of y_1 and y_2 in such a way that it minimizes his maximum expected payoff (loss). This can be done by plotting the expected payoff of player B for different strategies of player A. For plotting the expected payoffs, we draw two parallel lines with one unit apart and mark a scale on each of them. These two lines will represent the two available mixed strategies to player B. To draw $E_1(y)$, we join a_{11} on scale II to a_{12} on scale I. Similarly, the remaining expected payoffs can be plotted on the same graph. Now identify the lowest point on the upper boundary region of these lines, which will give the minimum expected payoff in the maximum expected payoff region on the upper boundary. This point will be the minimax value for player B and thus will determine the optimal values of y_1 and y_2. Now, the optimal strategies of player A will be determined with the help of corresponding lines which pass through this minimax point. Finally, the dimension of the payoff matrix will be reduced from $m \times 2$ to 2×2, which can be easily solved using the method discussed earlier.

Almost in the same manner, a $2 \times n$ payoff matrix can also be reduced into a 2×2 payoff matrix. In this case, instead of looking at the lowest point in the upper boundary region of the graph of the expected payoffs (A's expected payoff), we will look for the highest point in the lower boundary region, which will give the maximum expected payoff in the minimum expected payoff region.

Example 9.3. *A soft drink company calculated the market share of two products against its major competitor having three products and found out the impact of additional advertisement for any one of its products against the other. The payoff matrix is as follows:*

$$
\begin{array}{cccc}
 & & \text{Competitor} & \\
 & I & II & III \\
\text{Company} \quad \begin{array}{c} I \\ II \end{array} & \begin{array}{c} 6 \\ 20 \end{array} & \begin{array}{c} 7 \\ 12 \end{array} & \begin{array}{c} 15 \\ 10 \end{array}
\end{array}
$$

What is the best strategy of the company as well as the competitor? What is the payoff obtained by the company and the competitor in the long run?

Solution: For the given payoff,

Minimax = 12 and Maximin = 10. Since the Minimax is not equal to the Maximin, the given game doesn't possess a saddle point. Let x_1 and x_2, (where $x_1 + x_2 = 1$) be the probabilities of the company doing additional campaigning for its first and second product, respectively. Then the company's expected payoffs, against the competitor's pure moves, are given by:

Competitor's moves	Expected payoff to the company
I	$E_1(x) = 6x_1 + 20x_2 = 6x_1 + 20(1 - x_1) = 20 - 14x_1$
II	$E_2(x) = 7x_1 + 12x_2 = 7x_1 + 12(1 - x_1) = 12 - 5x_1$
III	$E_3(x) = 15x_1 + 10x_2 = 15x_1 + 10(1 - x_1) = 10 + 5x_1$

Here, $E_i(x)$, $i = 1, 2, 3$ denotes the company's expected payoff when its competitor uses its first, second, and third product, respectively. Now we plot these expected payoffs as shown in the following figure.

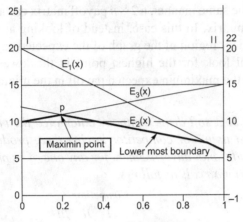

Fig. 9.1

The two parallel lines are at a unit distance apart and denote the two given strategies (products) for the company. To plot $E_1(x)$, we join 6 on II to 20 on I. Similarly, the other two expected payoffs of the company are plotted. The bold lines in the graph represent the lowermost envelope of the graph, which in fact represents the worst possible outcomes to the company. Now according to the maximin principle, the company will select the best of these worst possible outcomes, and this is given by point p in the graph, and this point denotes the value of the company.

The point identifies the two best strategies, II and III, of the competitor. Thus, the competitor will never play its first strategy, and the reduced payoff matrix is given as follows:

Competitor

II   III

Company   I   7   15
               11   12   10

Let y_2 and y_3 (where $y_2 + y_3 = 1$) denote the probabilities of the competitor doing additional campaigning for its second and third products respectively. Here $y_1 = 0$. Now, using the formula given in theorem 9.2, we have:

$$x_1 = \frac{a_{22} - a_{21}}{a_{11} + a_{22} - (a_{12} + a_{21})}$$

$$= \frac{10 - 12}{7 + 10 - (15 + 12)} = 0.2$$

and $\qquad x_2 = 1 - x_1 = 1 - 0.2 = 0.8.$

Similarly, $\qquad y_2 = \dfrac{a_{22} - a_{12}}{a_{11} + a_{22} - \left(a_{12} + a_{21}\right)} = \dfrac{-5}{-10} = 0.5$

and $\qquad y_3 = 1 - y_2 = 0.5$

and the value of the game is,

$$v = \frac{a_{11}a_{22} - a_{21}a_{12}}{a_{11} + a_{22} - \left(a_{12} + a_{21}\right)}$$

$$= \frac{7*10 - 12*15}{-10} = \frac{-110}{-10} = 11.$$

Hence, the solution of the game is:

(*a*) The optimal mixed strategies of the company are (0.2, 0.8).

(*b*) The optimal mixed strategies of the competitor are (0, 0.5, 0.5).

(*c*) The expected value of the game is $v = 11$, which denotes the payoff obtained by the company in the long run, and $v = -11$ will be the payoff obtained by the competitor in the long run.

Example 9.4. *Use the graphical method to find the solution of the following game problems.*

(*i*)

$$\begin{array}{c} & B \\ & \begin{array}{cc} I & II \end{array} \\ \begin{array}{c} I \\ II \\ A \ III \\ IV \\ V \end{array} & \begin{pmatrix} 1 & 2 \\ 4 & 1 \\ 4 & 2 \\ 5 & 3 \\ 2 & 4 \end{pmatrix} \end{array}$$

(*ii*)

$$\begin{array}{c} & B \\ & \begin{array}{cccc} I & II & III & IV \end{array} \\ A \ \begin{array}{c} I \\ I \end{array} & \begin{pmatrix} -1 & 2 & 0 & 4 \\ 4 & 5 & 6 & 3 \end{pmatrix} \end{array}$$

(*i*) **Solution:** Clearly the given game problem does not possess a saddle point. Let y_1 and y_2 denote the probabilities of player B playing his strategies

I and II, respectively, where $y_1 + y_2 = 1$, and then for the expected payoff to player B when player A plays his different strategies, we have:

Strategies of player A	Expected payoff to player B
I	$E_1(y) = y_1 + 2y_2 = y_1 + 2(1 - y_1) = 2 - y_1$
II	$E_2(y) = 4y_1 + y_2 = 4y_1 + (1 - y_1) = 1 + 3y_1$
III	$E_3(y) = 4y_1 + 2y_2 = 4y_1 + 2(1 - y_1) = 2 + 2y_1$
IV	$E_4(y) = 5y_1 + 3y_2 = 5y_1 + 3(1 - y_1) = 3 + 2y_1$
V	$E_5(y) = 2y_1 + 4y_2 = 2y_1 + 4(1 - y_1) = 4 - 2y_1$

The two parallel lines are at a unit distance apart and denote the two given strategies of player B. To plot E_1, we join 2 on scale I to 1 on scale II. Similarly, the remaining expected payoffs of player B corresponding to the other strategies of A are also plotted. The bold lines in the graph represent the uppermost boundary envelope of the graph, which in fact represent the worst possible outcomes to player B (minimizing player), since this region corresponds to maximum payoff (gain) to player A and hence maximum loss to B. Now according to the minimax principle, the company will select the best of these worst possible outcomes, which will result in a minimum loss to the company. This point which minimizes its maximum loss is denoted by p in the graph, and thus it is the minimax value for player B.

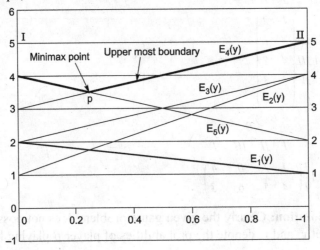

Fig. 9.2

The minimax point corresponds to the two best strategies, IV and V, of player A. Thus, player A will never play his strategies I, II, and III, and thus the reduced payoff matrix is:

$$B$$

		I	II
A	IV	5	3
	V	2	4

Let x_4 and x_5 (where $x_4 + x_5 = 1$) denote the probabilities of playing the strategies IV and V by player A, respectively. Here, $x_1 = x_2 = x_3 = 0$.

Now, using the formula given in theorem 9.2, we have:

$$x_4 = \frac{a_{22} - a_{21}}{a_{11} + a_{22} - (a_{12} + a_{21})} = \frac{4-2}{5+4-(3+2)} = 0.5$$

and

$$x_5 = 1 - x_4 = 1 - 0.5 = 0.5.$$

Similarly,

$$y_1 = \frac{a_{22} - a_{12}}{a_{11} + a_{22} - (a_{12} + a_{21})} = \frac{1}{4} = 0.25,$$

$$y_2 = 1 - y_1 = 0.75$$

and the value of the game is,

$$v = \frac{a_{11}a_{22} - a_{21}a_{12}}{a_{11} + a_{22} - (a_{12} + a_{21})}$$

$$= \frac{5*4 - 2*3}{4} = \frac{14}{4} = 2.5.$$

Hence, the solution of the game is:

(*a*) The optimal mixed strategy of player A is (0 0 0 0.5 0.5).

(*b*) The optimal mixed strategy of player B is (0.25 0.75).

(*c*) The expected value of the game is $v = 2.5$.

(*ii*) The game does not possess a saddle point. Let x_1 and x_2 (where $x_1 + x_2 = 1$) be the probabilities of player A using his first and second strategies respectively. The expected payoffs of player A is plotted as shown in the following figure below:

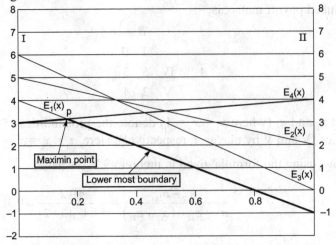

Fig. 9.3

The lowermost region of the graph represents the worst possible outcomes for the player A, since this envelope corresponds to the minimum payoff to player A, and A has to find a point where he will get a maximum payoff in this region. The highest point of this boundary is at the point P at which he will have the maximum of his minimum payoff. This point represents the maximin value for player A and identifies the two best strategies for player B, namely, I and IV. Thus, the reduced payoff matrix is:

$$
\begin{array}{cc}
 & \text{I} \quad \text{IV} \\
A & \begin{array}{c} \text{I} \\ \text{II} \end{array}\begin{pmatrix} -1 & 4 \\ 4 & 3 \end{pmatrix}
\end{array}
$$

Let y_1 and y_4 (where $y_1 + y_4 = 1$) denote the probabilities of playing I and IV strategies, respectively, by player B. Here, $y_2 = y_3 = 0$.

Now, using the formula given in theorem 9.2, we have:

$$
x_1 = \frac{a_{22} - a_{21}}{a_{11} + a_{22} - (a_{12} + a_{21})}
$$

$$
= \frac{3-4}{-1+3-(4+4)} = 0.16667
$$

and $\quad x_2 = 1 - x_1 = 1 - 0.16667 = 0.8333.$

Similarly, $\quad y_1 = \dfrac{a_{22} - a_{12}}{a_{11} + a_{22} - (a_{12} + a_{21})} = \dfrac{-1}{-6} = 0.16667,$

$$y_4 = 1 - y_1 = 0.8333$$

The value of the game is,

$$v = \dfrac{a_{11}a_{22} - a_{21}a_{12}}{a_{11} + a_{22} - (a_{12} + a_{21})} = \dfrac{-3-16}{-6} = 3.6667.$$

Hence, the solution of the game is:

(*a*) The optimal mixed strategy of player A is (0.16667 0.8333).

(*b*) The optimal mixed strategy of player B is (0.16667 0 0 0.8333).

(*c*) The expected value of the game is $v = 3.6667$.

9.7. Method of Dominance

In the previous section, we have discussed the method of reducing a $2 \times n$ or $m \times 2$ payoff matrix into a 2×2 payoff matrix using the graphical method. Here in this section, we will discuss how the size of an $m \times n$ payoff matrix can be reduced to a 2×2 by eliminating a course of action which is inferior to another, as this course of action will never be used by the player, and this procedure is known as the method of dominance. The following are the rules involved in this method:

(*i*) If all the elements in the ith row are less than or equal to the corresponding elements of another row, say the jth row, then the ith will be said to be dominated by the jth row. In this case, the row player (maximizing player) will never be in a better position by playing the ith row (or the ith strategy), and hence the ith row will be deleted from further consideration from the given payoff table.

(*ii*) Similarly, if all the elements of the ith column are greater than or equal to the corresponding elements of another column, say the jth column, then the ith column is said to be dominated by the jth column. In this case, the column player (minimizing player) whose objective is to minimize his loss will never play the ith column (or the ith strategy), since it will give a better payoff to his opponent, and hence the ith column will be deleted from further consideration from the given payoff matrix.

(*iii*) A pure strategy may also be dominated by the average of two or more other pure strategies. If all the elements of the average of a number of rows are greater than or equal to the corresponding elements of another row, say the ith, then the average will dominate the ith row, and hence the ith row will be eliminated from the payoff matrix. Similarly, in the case of the columns, if the average of two or more columns is less than or equal to the corresponding elements of another column, say the jth column, then the jth *column* is said to be dominated by the average of a number of columns and hence can be eliminated from the payoff table.

Example 9.5. *Use dominance method to solve the following game.*

$$
(i) \quad \begin{array}{c} \\ A \end{array} \begin{array}{c} I \\ II \\ III \end{array} \overset{\begin{array}{c} B \\ I \quad II \quad III \end{array}}{\begin{pmatrix} 10 & 5 & 3 \\ 4 & 5 & 2 \\ 5 & 3 & 6 \end{pmatrix}}
$$

$$
(ii) \quad \begin{array}{c} \\ A \end{array} \begin{array}{c} I \\ II \\ III \end{array} \overset{\begin{array}{c} B \\ I \quad II \quad III \quad IV \end{array}}{\begin{pmatrix} 1 & 5 & 3 & 4 \\ 3 & 1 & 2 & 6 \\ 5 & 3 & 6 & 5 \end{pmatrix}}
$$

(*i*) **Solution:** In the given payoff matrix, it can be clearly seen that every element of the first row is greater than or equal to the corresponding elements of the second row. So, from player A's point of view, the second strategy is dominated by the first strategy or the first strategy dominates the second strategy. Hence, player A will never play his second strategy, and thus the second row may be deleted. The payoff matrix is reduced to the form:

$$
\begin{array}{c} \\ A \end{array} \begin{array}{c} I \\ III \end{array} \overset{\begin{array}{c} B \\ I \quad II \quad III \end{array}}{\begin{pmatrix} 10 & 5 & 3 \\ 5 & 3 & 6 \end{pmatrix}}
$$

In the reduced payoff, from the player B point of view, his II strategy dominates I, since every element in the second column is less than or equal to the corresponding elements of the first column. Hence, the first column can now be deleted; the following is the reduced payoff matrix:

$$B$$

$$
A \begin{array}{c} \\ \text{I} \\ \text{III} \end{array}
\begin{array}{cc} \text{II} & \text{III} \\ \begin{pmatrix} 5 & 3 \\ 3 & 6 \end{pmatrix} \end{array}
$$

Now, let x_1 and x_3 (where $x_1 + x_3 = 1$) denote the probabilities of player A using his I and III strategies, respectively, and let y_2 and y_3 (where $y_2 + y_3 = 1$) denote the probabilities of player B playing his strategies II and III, respectively.

Here x_2 = probability of playing strategy II by player A = 0 and y_1 = probability of playing strategy I by player B = 0

Now the reduced 2×2 game problem can be solved as follows:

$$
x_1 = \frac{a_{22} - a_{21}}{a_{11} + a_{22} - (a_{12} + a_{21})} = \frac{6-3}{5+6-(3+3)} = 0.6,
$$

and $\qquad x_3 = 1 - x_1 = 1 - 0.6 = 0.4.$

Similarly, $\qquad y_2 = \dfrac{a_{22} - a_{21}}{a_{11} + a_{22} - (a_{12} + a_{21})} = \dfrac{3}{5} = 0.6.$

and $\qquad y_3 = 1 - y_2 = 0.4.$

The value of the game is,

$$
v = \frac{a_{11} a_{22} - a_{21} a_{12}}{a_{11} + a_{22} - (a_{12} + a_{21})} = \frac{30-9}{5} = 4.2.
$$

Thus, the solution of the game is

(a) The optimal mixed strategy of player A is (0.6 0 0.4).

(b) The optimal mixed strategy of player B is (0 0.6 0.4).

(c) The expected value of the game is v = 4.2.

(ii) The first column dominates the fourth column since every element of the first column is less than or equal to the corresponding element of the fourth column. Hence, player B will never play his fourth strategy, and the reduced payoff is;

$$B$$

$$
A \begin{array}{c} \\ \text{I} \\ \text{II} \\ \text{III} \end{array}
\begin{array}{ccc} \text{I} & \text{II} & \text{III} \\ \begin{pmatrix} 1 & 5 & 3 \\ 3 & 1 & 2 \\ 5 & 3 & 6 \end{pmatrix} \end{array}
$$

Now, neither the column nor the row dominance property can be applied to the reduced matrix, so check for average domination. The average of the first and second column is (3, 2, 4), and these elements are less than or equal to the corresponding elements of the third column, and hence the average will dominate the third column. The third column can now be deleted, and the reduced payoff is:

$$
A \begin{array}{c} \\ \text{I} \\ \text{II} \\ \text{III} \end{array}
\begin{array}{c} B \\ \begin{array}{cc} \text{I} & \text{II} \end{array} \\ \begin{pmatrix} 1 & 5 \\ 3 & 1 \\ 5 & 3 \end{pmatrix} \end{array}
$$

In the reduced matrix, the third row dominates the second row. The second row can now be deleted and the reduced payoff is:

$$
A \begin{array}{c} \\ \text{I} \\ \text{III} \end{array}
\begin{array}{c} B \\ \begin{array}{cc} \text{I} & \text{II} \end{array} \\ \begin{pmatrix} 1 & 5 \\ 5 & 3 \end{pmatrix} \end{array}
$$

The reduced 2 × 2 game problem can be solved as follows:

Let x_1 and x_3 (where $x_1 + x_3 = 1$) denote the probabilities of player A using his I and III strategies respectively, and let y_1 and y_2 (where $y_1 + y_2 = 1$) be the probabilities of player B playing his strategies I and II, respectively. We have $x_2 = 0, y_3 = y_4 = 0$:

$$
x_1 = \frac{a_{22} - a_{21}}{a_{11} + a_{22} - (a_{12} + a_{21})} = \frac{3 - 5}{1 + 3 - (5 + 5)} = 0.3333
$$

and

$$
x_3 = 1 - x_1 = 1 - 0.3333 = 0.6667.
$$

Similarly,

$$
y_1 = \frac{a_{22} - a_{21}}{a_{11} + a_{22} - (a_{12} + a_{21})} = \frac{-2}{-6} = 0.3333,
$$

and

$y_2 = 1 - y_1 = 0.6667$ and the value of the game is,

$$
v = \frac{a_{11}a_{22} - a_{21}a_{12}}{a_{11} + a_{22} - (a_{12} + a_{21})} = \frac{3 - 25}{-6} = 3.6667.
$$

Hence, the solution of the game is:

(a) The optimal mixed strategy of player A is (0.3333 0 0.6667).

(b) The optimal mixed strategy of player B is (0.3333 0.6667 0 0).

(c) The expected value of the game is $v = 3.6667$.

9.8. Solution of a Game Using the Simplex Method

This method is useful for solving any rectangular game with an $m \times n$ payoff matrix if it does not have a saddle point and at the same time cannot be simplified by the graphical or dominance method. In this method, we will convert the given game into a linear programming problem and solve it by using the simplex method.

Let us consider an $m \times n$ game problem with $A = (a_{ij})$ as its payoff matrix. Also let $x = (x_1, ..., x_2, x_m)$ and $y = (y_1, ..., y_2, y_n)$ be the probabilities of playing different strategies by the maximizing (row) and minimizing (column) player, respectively, where $0 \leq x_i, y_j \leq 1$, $\sum_{i=1}^{m} x_i = 1$, and $\sum_{j=1}^{n} y_j = 1$. Here, m and n are the strategies available to player 1 and 2, respectively.

Let E_j denote the expected gain to the maximizing player when the minimizing player plays his j^{th} strategy, and then:

$$\sum_{i=1}^{m} a_{i1} x_i = E_1 \text{, when minimizing player plays his first strategy}$$

$$\sum_{i=1}^{m} a_{i2} x_i = E_2 \text{, when minimizing player plays his second strategy}$$

$$\vdots$$

$$\sum_{i=1}^{m} a_{in} x_i = E_n \text{, when minimizing player plays his } n^{th} \text{ strategy}$$

The objective of the maximizing player is to select x_i such that he can maximize his minimum expected gains. Let u denote the minimum expected gain of the maximizing player, where $u = \text{Min}\left[\sum_{i=1}^{m} a_{ij} x_i, j = 1, 2, ..., n.\right]$. We have

$$Max\, u = Min\frac{1}{u} = Min \sum_{i=1}^{m} \frac{x_i}{u} \quad \left(\text{Since } \sum_{i=1}^{m} x_i = 1\right)$$

Subject to the constraints

$$\sum_{i=1}^{m} a_{ij} x_i \geq u \text{ , } \forall j \text{ and}$$

$$\sum_{i=1}^{m} x_i = 1, \; x_i \geq 0, \; i = 1, 2... m$$

Assuming $u > 0$ and letting $x_i' = \frac{x_i}{u}$, we have

$$Min\, u' = \left(\frac{1}{u}\right) = \sum_{i=1}^{m} x_i'$$

Subject to

$$\sum_{i=1}^{m} a_{ij} x_i' \geq 1, \quad j = 1, 2, \ldots, n$$

$$x_i' \geq 0, \qquad i = 1, 2, \ldots, m$$

Similarly, the minimizing player will try to minimize his maximum losses, and the resulting problem can be written as:

$$Min\, v = Max\frac{1}{v} = \sum_{j=1}^{n} y_j'$$

Subject to

$$\sum_{j=1}^{n} a_{ij} y_j' \leq 1 \quad i = 1, 2, \ldots, m$$

$$y_j' \geq 0, \qquad j = 1, 2, \ldots, n$$

Where $y_j' = \dfrac{y_j}{v}$ and $v = Max\left[\sum_{j=1}^{n} a_{ij} y_j, j = 1, 2, \ldots, n\right]$ is the maximum expected loss of the minimizing player when the other player plays his various strategies.

Remarks

1. A linear programming problem requires all the variables involved in the problem to be non-negative, and thus with a non-negative value of the game. This can be assured by making all the elements of the given payoff matrix to be greater than or equal to zero. If there is any negative element in the payoff table, then a large constant can be added to all the elements of the matrix so that the minimum element in the matrix is greater than or equal to zero. The optimal solution in terms of probabilities of choosing various strategies for the new problem will always be the same as it was for the original problem, but the value of the game for the original problem will be equal to the value of game for the new problem minus the constant (which was added initially in the payoff matrix).

2. It can be easily seen that the two problems are dual to one another, and hence all the rules of duality can be applied here to get the optimal strategies of both the players.

3. It will be more economical to solve the LP formulation of the minimizing player than the maximizing player (since no artificial variable will be needed to solve this problem).

Example 9.6. *Give the LPP formulation of the game given in Examples 9.4 (ii) and 9.5 (i) and solve it using the simplex method.*

(i)

$$A \begin{array}{c} \\ I \\ I \end{array} \begin{array}{cccc} & I & II & III & IV \\ \begin{pmatrix} -1 & 2 & 0 & 4 \\ 4 & 5 & 6 & 3 \end{pmatrix} \end{array}$$

(ii)

$$A \begin{array}{c} \\ I \\ II \\ III \end{array} \begin{array}{ccc} & I & II & III \\ \begin{pmatrix} 10 & 5 & 3 \\ 4 & 5 & 2 \\ 5 & 3 & 6 \end{pmatrix} \end{array}$$

(i) Since the given payoff matrix has a negative element, add a suitable positive number, say 1, to all the elements of the matrix, so that each element of the given payoff matrix becomes greater than or equal to zero. Now the resulting payoff matrix is given as follows:

$$A \begin{array}{c} \\ I \\ II \end{array} \begin{array}{cccc} I & II & III & IV \\ \begin{pmatrix} 0 & 3 & 1 & 5 \\ 5 & 6 & 7 & 4 \end{pmatrix} \end{array}$$

Let $x = (x_1, x_2)$ and $y = (y_1, y_2, y_3, y_4)$ be the probabilities associated with the strategies of the players A and B, respectively, where $0 \leq x_i, y_j \leq 1 \ \forall, i, j$. Further $\sum_{i=1}^{2} x_i = 1$ and $\sum_{j=1}^{4} y_j = 1$. To find the optimal strategies of player A, the linear programming problem can be formulated as:

$$\text{Max } u = \text{Min} \frac{1}{u} = x_1' + x_2'$$

Subject to

$$0x_1' + 5x_2' \geq 1$$
$$3x_1' + 6x_2' \geq 1$$
$$x_1' + 7x_2' \geq 1$$
$$5x_1' + 4x_2' \geq 1$$
$$x_1', x_2' \geq 0,$$

where $x_1' = \dfrac{x_i}{u}$, $i = $ i, 2 and u is the expected minimum gain to player A.

Similarly, linear programming formulation of the game for player B is

$$\text{Min } v = \text{Max } \frac{1}{v} = y_1' + y_2' + y_3' + y_4'$$

Subject to

$$0y_1' + 3y_2' + y_3' + 5y_4' \leq 1$$
$$5y_1' + 6y_2' + 7y_3' + 4y_4' \leq 1$$
$$y_1', y_2', y_3', y_4' \geq 0,$$

where $y_j' = \dfrac{y_j}{v}$, $j = 1, 2, 3, 4$ and v is the expected maximum loss to player B.

The canonical form to B's problem is,

$$0y_1' + 3y_2' + y_3' + 5y_4' + s_1 = 1$$
$$5y_1' + 6y_2' + 7y_3' + 4y_4' + s_2 = 1$$
$$\frac{1}{v} - y_1' - y_2' - y_3' - y_4' = 0$$
$$y_1', y_2', y_3', y_4', s_1, s_2, s_3 \geq 0,$$

Initial Table

Basic Variables	y_1'	y_2'	y_3'	y_4'	s_1	s_2	solution
s_1	0	3	1	5	1	0	1
s_2	5	6	7	4	0	1	1
$\dfrac{1}{v}$	−1	−1	−1	−1	0	0	0

And the corresponding optimal table is:

Basic Variables	y_1'	y_2'	y_3'	y_4'	s_1	s_2	solution
y_4'	0	0.0600	0.200	1	0.200	0	0.200
y_1'	1	0.720	1.240	0	−0.160	0.200	0.040
$\dfrac{1}{v}$	0	0.320	0.440	0	0.040	0.200	0.240

$v = \dfrac{1}{0.240} = 4.1667$ and hence, the expected value of the game is 4.1667

$-1 = 3.1667$.

The optimum strategies for player B are $y_1 = y_1' * 4.1667, = 0.1667, y_2 = 0,$ $y_3 = 0$, and $y_4 = y_4' * 4.16667 = 0.8333$

Using the duality theorem, from the previous optimal table of B's problem, we can obtain the optimum solution to A's problem, that is, we have $x_1' = 0.040$ and $x_2' = 0.200$. Further, $u = (1 / (x_1' + x_2')) = (1 / 0.24) = 4.1667 = v.$

Therefore, the optimum strategies for player A are $x_1 = 0.040 * 4.1667 = 0.1667$ and $x_2 = 0.200 * 401667 = 0.8333$. (These solutions are in agreement with the solution of the same problem obtained using the graphical method.)

(*ii*) **Solution:** Let $x = (x_1, x_2, x_3)$ and $y = (y_1, y_2, y_3)$ be the probabilities associated with the strategies of player A and B, respectively, where $0 \le x_i, y_j \le 1, \forall i, j.$

Further, $\sum_{i=1}^{3} x_i = 1$ and $\sum_{j=1}^{3} y_j = 1$.

Linear programming formulation of the game for player A is:

$$\text{Max } u = \text{Min } u = x_1' + x_2' + x_3'$$

Subject to

$$10x_1' + 4x_2' + 5x_3' \ge 1$$
$$5x_1' + 5x_2' + 3x_3' \ge 1$$
$$3x_1' + 2x_2' + 3x_3' \ge 1$$
$$x_1', x_2', x_3' \ge 0,$$

where $x_1' = \dfrac{x_i}{u}$, $i = 1, 2, 3$ and u is the is the expected minimum gain to player A.

Similarly, linear programming formulation of the game for player B is:

$$\text{Min } v = \text{Min } \frac{1}{v} = y_1' + y_2' + y_3'$$

Subject to

$$10y_1' + 5y_2' + 3y_3' \le 1$$
$$4y_1' + 5y_2' + 2y_3' \le 1$$
$$5y_1' + 3y_2' + 6y_3' \le 1$$
$$y_1', y_2', y_3' \ge 0$$

where $y_j' = \dfrac{y_j}{v}$, $j = 1, 2, 3$ and v is the expected maximum loss to player B.

The canonical form to B's problem is:

$$10y_1' + 5y_2' + 3y_3' + s_1 = 1$$
$$4y_1' + 5y_2' + 2y_3' + s_2 = 1$$
$$5y_1' + 3y_2' + 6y_4' + s_3 = 1$$
$$\frac{1}{v} - y_1' - y_2' - y_3' = 0$$
$$y_1', y_2', y_3', s_1, s_2, s_3 \geq 0,$$

Initial Table

Basic Variables	y_1'	y_2'	y_3'	s_1	s_2	s_3	Solution
s_1	10	5	3	1	0	0	1
s_2	4	5	2	0	1	0	1
s_3	5	3	6	0	0	1	1
$\dfrac{1}{v}$	−1	−1	−1	0	0	0	0

And the corresponding optimal table is:

Basic Variables	y_1'	y_2'	y_3'	s_1	s_2	s_3	Solution
y_2'	2.1429	1	0	0.2857	0	−0.1429	0.1429
s_2	−6.2381	0	0	−1.1429	1	0.2381	0.0952
y_3'	−0.2381	0	1	−0.1429	0	0.2381	0.0952
$\dfrac{1}{v}$	0.9048	0	0	0.1429	0	0.0952	0.2381

Hence, the expected value of the game is $v = \dfrac{1}{0.2381} = 4.2$.

The optimum strategies for player B are $y_1 = 0$, $y_2 = y_2' * 4.2 = 0.1429 * 4.2 = 0.6$, and $y_3 = y_3' * 4.2 = 0.0952 * 4.2 = 0.4$.

Using the duality theorem, from the previous optimal table of B's problem, we the optimum solution to A's problem: $x_1' = 0.1429$, $x_2' = 0$, and $x_3' = 0.0952$. Also, we have $u = (1/0.2381) = 4.2 = v$.

Therefore, the optimum strategies for player A are $x_1 = 0.1429 * 4.2 = 0.6$, $x_2 = 0$, and $x_3 = 0.0952 * 4.2 = 0.4$. (These solutions are in agreement with the solution of the same problem obtained using the dominance method).

9.9. Solution of a Game Using Gambit

Gambit is an open-source collection of tools for doing computation in game theory. Using Gambit, we can easily build, analyze, and explore given game models. The following link can be used for downloading and installing Gambit:

http://gambit.sourceforge.net

We use (*ii*) of Example 9.6 to understand how Gambit can be use to find the optimal strategies of a given game problem:

Step 1: Go to All Programs and open Gambit. Go to the File option in the menu bar and select **New → Strategic Game** as shown in the following figure:

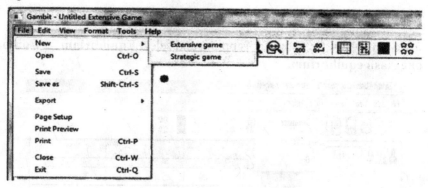

Fig. 9.4

Step 2: On executing the previous steps, a new window will open now. By default only two strategies are shown for both the players. We can increase the number of strategies as we wish by clicking on **"Add a strategy for this player"** for both the players.

Gambit - Untitled Strategic Game (unsaved changes)				
Player A		1		2
Add a strategy for this player	1	0	0	0	0
	2	0	0	0	0

Fig. 9.5

Step 3: Enter the payoff matrix. Every cell of the payof matrix will have two values, one for Player A and the other for Player B. In the case of a zero-sum game, the Player B entry will be equal to the negative of the Player A entry as shown in the following figure:

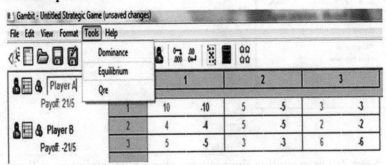

Fig. 9.6

Step 4. Now go to the Tools option and select **Equilibrium → Compute one Nash equilibrium.**

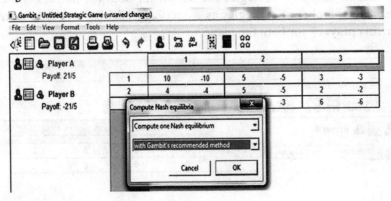

Fig. 9.7

Fig. 9.8

Step 5. Now click OK and we will have the following optimal solution:

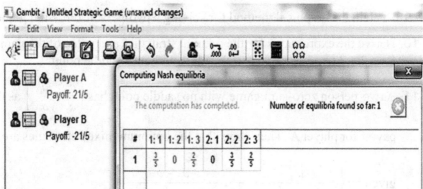

Fig. 9.9

This implies:

(*a*) The optimal strategy of Player A is $\left(\dfrac{3}{5}, 0, \dfrac{2}{5}\right)$.

(*b*) The optimal strategy of Player B is $\left(0, \dfrac{3}{5}, \dfrac{2}{5}\right)$.

EXERCISES

1. What is game theory? Explain the role of the theory of games in decision making.

2. What is a rectangular game? What are pure strategy and mixed strategy in a game?

3. Explain the following terms: Pure strategy, Mixed strategy, Saddle point, Zero-sum game, Fair game, and Payoff matrix.

4. Explain the maxmin and minimax principle used in game theory.

5. Define a saddle point. Explain the steps for determining the saddle point of a game.

6. What are the various characteristics of a competitive situation to be called a competitive game?

7. Explain the various methods for solving a two-person zero-sum game.

8. Discuss the dominance method for solving a two-person zero-sum game.

9. Explain the graphical method for solving a two-person zero-sum game.

10. Derive the expressions for the optimal strategies and value of the game for a 2×2 two-person zero-sum game without a saddle point.

11. A two-person zero-sum game with no saddle point has $\begin{pmatrix} a_{11} & a_{12} \\ a_{21} & a_{22} \end{pmatrix}$ as the

payoff for player A. Then show that the optimum mixed strategies are

given by $\left(\dfrac{a_{22} - a_{21}}{a_{11} + a_{22} - \left(a_{12} + a_{21}\right)}, \dfrac{a_{11} - a_{12}}{a_{11} + a_{22} - \left(a_{12} + a_{21}\right)} \right)$

$\left(\dfrac{a_{22} - a_{12}}{a_{11} + a_{22} - \left(a_{12} + a_{21}\right)}, \dfrac{a_{11} - a_{21}}{a_{11} + a_{22} - \left(a_{12} + a_{21}\right)} \right)$ and the expected gain for

player A is given by $v = \dfrac{a_{11}a_{22} - a_{21}a_{12}}{a_{11} + a_{22} - \left(a_{12} + a_{21}\right)}$.

12. Establish the relation between a two-person zero-sum game and a linear programming problem.

13. Explain the process of solving a two-person zero-sum game using the simplex method.

14. For any two-person zero-sum game with $(a_{ij})_{m \times n}$ as its payoff, show that the maximin value $\underline{v}$ of the matrix will always be less or equal to the minimax value $\bar{v}$ of the matrix; that is, $\max_i \min_j (a_{ij}) \le \min_j \max_i (a_{ij})$.

15. The following games are deterministic games. Determine the saddle point and optimum strategies for each player:

(i) $\begin{pmatrix} 8 & 2 \\ 4 & 3 \end{pmatrix}$
(ii) $\begin{pmatrix} 1 & -4 & 4 \\ 0 & 2 & 5 \\ 3 & 2 & 7 \end{pmatrix}$
(iii) $\begin{pmatrix} 10 & 25 & 14 \\ 21 & 19 & 15 \\ 16 & 18 & 12 \end{pmatrix}$

(iv) $\begin{pmatrix} 5 & 3 & 6 \\ 9 & 4 & 5 \end{pmatrix}$
(v) $\begin{pmatrix} -5 & -6 \\ -1 & -10 \\ -9 & -8 \end{pmatrix}$

16. Use the mehtod of dominance to solve the following game problems:

(*i*) $\begin{pmatrix} 10 & 5 & 1 \\ 21 & 9 & 5 \\ 15 & 8 & 10 \end{pmatrix}$

(*ii*) $\begin{pmatrix} 13 & 11 & 18 & 10 \\ 8 & 14 & 9 & 10 \\ 14 & 12 & 16 & 14 \end{pmatrix}$

(*iii*) $\begin{pmatrix} 1 & -2 & 0 \\ 4 & 0 & 3 \\ 1 & 2 & 1 \\ 5 & 0 & 1 \end{pmatrix}$

17. Use the graphical method to solve the following game problem:

(*i*) $\begin{pmatrix} 2 & 6 & 0 \\ 5 & 3 & 4 \end{pmatrix}$

(*ii*) $\begin{pmatrix} 1 & -4 \\ 2 & 2 \\ -1 & 3 \end{pmatrix}$

(*iii*) $\begin{pmatrix} 44 & 15 & 10 & 21 \\ 35 & 19 & 20 & 18 \end{pmatrix}$

18. Use the simplex method to solve the following game problems:

(*i*) $\begin{pmatrix} 6 & 7 \\ 4 & 5 \end{pmatrix}$

(*ii*) $\begin{pmatrix} 6 & -2 & 0 \\ 4 & 0 & 3 \\ 1 & 2 & 1 \\ 5 & 0 & 1 \end{pmatrix}$

(*iii*) $\begin{pmatrix} 1 & 4 \\ 2 & 2 \\ 0 & 3 \end{pmatrix}$

USE OF MATHEMATICA, MATLAB, LINDO, AND WINQSB TO SOLVE LINEAR PROGRAMMING MODELS

A.1. Linear Programming Problems using MATHEMATICA

The Wolfram Language has a collection of algorithms for solving linear optimization problems with real variables, accessed via LinearProgramming, NMinimize, NMaximize, Minimize, and Maximize. LinearProgramming gives direct access to linear programming algorithms, provides the most flexibility for specifying the methods used, and is the most efficient for large-scale problems. NMinimize, NMaximize, Minimize, and Maximize are convenient for solving linear programming problems in equation and inequality form.

(For more details, one may visit *https://reference.wolfram.com/language/tutorial/ConstrainedOptimizationLinearProgramming.html#430238072*)

Example A.1. *Min z = 2x₁ + 1.5x₂*

Subject to constraint

$$6x_1 + 3x_2 \geq 20$$
$$2x_1 + x_2 \, 2 \geq 15$$
$$x_1 + x_2 \geq 8$$
$$x_1, x_2 \geq 0$$

(*i*) Using **Linear Programming**

Linear Programming $[\{2, 1.5\}, \{\{6, 3\}, \{2, 1\}, \{1, 1\}\}, \{\{20, 1\}, \{15, 1\}, \{8, 1\}\}, \{\{0, \text{Infinity}\}, \{0, \text{Infinity}\}\}]$

Output: $\{7., 1.\}$, *i.e.* $x_1 = 7$ and $x_2 = 1$

The following is the screenshot of the previous calculation:

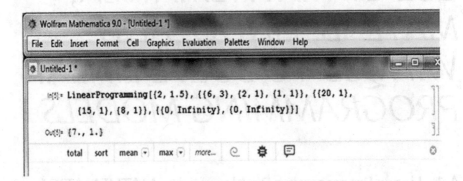

Remark: **The Linear Programming** function works with minimization as an objective function; if the problem is maximization, then it has to be converted to minimization to use **Linear Programming.** Also in this function equality relation will be denoted by 0, less than equality by -1, and greater than equality by 1.

(*ii*) Using **Minimize**

Minimize $[\{2x_1 + 1.5x_2, 6x_1 + 3x_2 \geq 20 \ \&\& \ 2x_1 + x_2 \geq 15 \ \&\& \ x_1 + x_2 \geq 8 \ \&\& \ x_1 \geq 0 \ \&\& \ x_2 \geq 0\}, \{x_1, x_2\}]$

Output: $\{15.5, \{x_1 \rightarrow 7., x_2 \rightarrow 1.\}\}$, *i.e.* $x_1 = 7$ and $x_2 = 1$ and $z = 15.5$

(*iii*) Using **NMinimize**

NMinimize $[\{2x_1 + 1.5x_2, 6x_1 + 3x_2 \geq 20 \ \&\& \ 2x_1 + x_2 \geq 15 \ \&\& \ x_1 + x_2 \geq 8 \ \&\& \ x_1 \geq 0 \ \&\& \ x_2 > 0\}, \{x_1, x_2\}]$

Output: $\{15.5, \{x_1 = 7., x_2 = 1.\}\}$, *i.e.* $x_1 = 7$ and $x_2 = 1$ and $z = 15.5$

Note:- Nminimize will convert the fractional solution into decimal form.

Example A.2. *Max z = 10x_1 + 6x_2 − 8x_3*

Subject to constraint

$$5x_1 + 2x_2 + 6x_3 \leq 20$$
$$10x_1 + 4x_2 - 6x_3 \leq 35$$
$$x_1, x_2, x_3 \geq 0$$

(*i*) Using **Linear Programming**

Linear Programming [{− 10, − 6, 8}, {{5, + 2, 6}, {10, 4, − 6}}, {{20, − 1}, {35, − 1}}, {{0, Infinity}, {0, Infinity}, {0,Infinity}}]

Output: $\left\{0, \dfrac{55}{2}, \dfrac{25}{3}\right\}$

(*ii*) Using **Maximize**

Maximize [{10 × 1 + 6x_2 − 8x_3, 5x_1 − 2x_2 + 6x_3 < = 20 && 10 × 1 + 4 × 2 − 6 × 3 < = 35 && × 1 > = 0 && x_2 > = 0 && x_3 > = 0}, {x_1, x_2, x_3}]

Output: $\left\{65, \{x_1 \to 0, x_2 \to \dfrac{55}{2}, x_3 \to \dfrac{25}{3}\}\right\}$

(*iii*) Using **NMaximize**

NMaximize [{10 × 1 + 6x_2 − 8x_3, 5x_1 − 2x_2 + 6x_3 < = 20 && 10 × 1 + 4 × 2 − 6 × 3 < = 35 && × 1 > = 0 && × 2 > = 0 && × 3 > = 0}, {x_1, x_2, x_3}]

Output: {65., {$x_1 \to 0.$, $x_2 \to 27.50$, $x_3 \to 12.50$}}

Duality using Mathematica: DualLinearProgramming function can be used to solve a primal-dual problem using **Mathematica.**

The following is the dual of the problem given in Example A.1.

$$Max\ w = 20w_1 + 15w_2 + 8w_3$$

Subject to constraint

$$6w_1 + 2w_2 + w_3 \leq 2$$
$$3w_1 + w_2 - w_3 \leq 1.5$$
$$w_1, w_2, w_3 \geq 0$$

DualLinearProgramming [{2, 1.5}, {{6, 3}, {2, 1}, {1, 1}}, {{20, 1}, {15, 1}, {8, 1}}, {{0, Infinity}, {0, Infinity}}]

Output. {{7., 1.}, {0., 0.5, 1.}, {0., 0}, {0., 0.}}

This implies and $x_1 = 7$, $x_2 = 1$ and $w_1 = 0$, $w_2 = 0.5$ and $w_3 = 1$.

A.2. Linear Programming Problems Using MATLAB

In MATLAB, the Optimization Toolbox provides functions for finding the solution that minimizes or maximizes objectives while satisfying constraints. This toolbox can be used to solve linear programming, mixed-integer linear programming, quadratic programming, nonlinear optimization, and nonlinear least squares. LPPs can be solved using the **linprog** function. The following are some of the syntax of **linprog**.

(For more details on this one may visit *http://in.mathworks.com/help/ optim/ug/linprog.html#buus0qo-2*)

1. *Min f*

Subject to

$$Ax \le b$$

Syntax for such a problem is **linprog** *(f, A, b)*.

Example A.3. *Min $z = x_1 + x_2$*

Subject to

$$3x_1 + x_2 \ge 6$$
$$x_1 + 5x_2 \ge 8$$

```
% Example Min z= x1+x2
% Subject to
%     3x1+x2>=6
%     x1+5x2>=8

%Co-eff matrix of the constraints
A =    [-3   -1;   -1   -5];
resource=[-6;   -8];
%co-eff of objective function
obcoeff = [1 1];
%solving
%solving
[x z]   = linprog  (obcoeff,  A,  resource)
```

Solution: $x_1 = 1.5714$, $x_2 = 1.2857$ and $z = 2.8571$

The following is the screenshot of the command (output) window of the previous calculation:

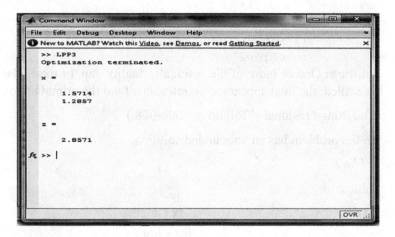

2. *Min f*

Subject to

$$\begin{cases} Ax \leq b \\ Aeq = beq \end{cases}$$

Syntax for such a problem is **linprog (f, A, b, Aeq, beq)**

Example A.4. *Min z = 2x$_1$ + 3x$_2$ + 4x$_3$*

Subject to

$$8x_1 + x_3 = 50$$

$$6x_2 + x_3 \leq 70$$

```
%Example Min z= 2x1+3x2+4x3
%Subject to
%      8x1+x3=50
%      6x2+x3?70
%
%Co-eff of inequality constraints
A =[0 6 1];
resource1=70;
%Co-eff of equality constraint
Aeq =[8 0 1];
```

```
resource2 =50;
%co-eff of objective function
object =   [2 3 4];
%solving
[x z]   = linprog(object, A,   resource1,  Aeq,
           resource2);
```

Solution: One or more of the residuals, duality gap, or total relative error has stalled: the dual appears to be infeasible (and the primal unbounded).

(The primal residual < TolFun = 1.00e-008.)

i.e. the problem has an unbounded solution.

3. *Min f*

Subject to

$$\begin{cases} Ax \leq b \\ Aeq = beq \end{cases}$$

$$lb \leq x \leq ub$$

Syntax for such a problem is linprog (*f, A, b, Aeq, beq, lb, ub*)

Example A.5. *Min z = 2x$_1$ + 8x$_2$*

Subject to

$$5x_1 + 10x_2 = 150$$

$$x_1 \leq 20$$

$$x_2 \geq 15$$

$$x_1, x_2 \geq 0$$

```
% Example      Min z=2x1+8x2
% Subject to constraint
% 5x1+10x2=150
% x1<=20
% x2>=15
% x1,x2 ≥0

%bounds of variables
lb =   [0 15];
ub =   [20 inf];
% co-efficient matrix for eqality constraints
```

```
eqconst =   [5 10];
resource2 = 150;
%objective function
obcoeff = [2 8];
%solving
warning ' off'
[x z]= linprog(obcoeff,[],[],eqconst,resource2,
        lb,ub)
```

Solution: $x_1 = 0$, $x_2 = 15$, and $z = 120$

Example A.6. *Min $z = 2x_1 + 1.5x_2$*

Subject to

$$6x_1 + 3x_2 \geq 20$$

$$2x_1 + x_2 \geq 15$$

$$x_1 + x_2 \geq 8$$

$$x_1, x_2 \geq 0$$

```
%   Min z=2x1+1.5x2
% Subject to constraint
%       6x1+3x2>=20
%       2x1+x2>=15
%       x1+x2>=8
%       x1,x2>=0

%bounds of variables
lb = zeros    (2,1);
%co-efficient matrix for inequality constraints
inconst =    [-6 -3;  -2 -1;  -1 -1];
resource1 =    [-20;  -15;  -8];
%objective function
obcoeff =   [2 1.5];
warning 'off'
%solving
[x z]= linprog(obcoeff,inconst,resource1,[],[],
        lb)
```

Solution: $x_1 = 7$, $x_2 = 1$, and $z = 15.5$

Example A.7. *Min z = $2x_1 - 3x_2 + 4x_3$*

Subject to

$$8x_1 + x_3 = 50$$

$$6x_2 + x_3 \leq 70$$

$$x_1, x_2 \geq 0 \text{ and } -15 \leq x_3 \infty$$

```
%Example Min z = 2x1 - 3x2 + 4x3
%Subject to
%        8x1-x3=50
%        6x2+x3<=70
%        x1,x2>=0        and        -15<=x3<=∞
%Co-eff of inequality constraints
A =[0 6 1];
resource1=70;
%bounds
lb =zeros(3,1);
ub =inf(3,1);
lb(3)=-15;
%Co-eff of equality constraint
Aeq =[8 0 -1];
resource2 =50;
%co-eff of objective function
object =[2  -3 4];
%solving
[x z]= linprog(object,A,resource1,Aeq,resource2
       ,lb,ub);
```

Solution: $x_1 = 4.375$, $x_2 = 14.1667$, $x_3 = -15.00$, and $z = -93.75$

A.3. Linear Programming Problems Using LINDO

With the LINDO API, we can easily create our own optimization applications. It also allows us to plug the power of the LINDO solver right into
customized applications that we have written. It has a comprehensive tool
designed to help us build and solve a wide range of optimization problems,

including linear programs, mixed integer programs, quadratic programs, and general nonlinear non-convex programs, and so forth.

When we start LINDO, our screen should resemble the following:

Initially we will see two windows, the outer window labeled LINDO is the main frame window. The main frame window also contains all the command menus and the command toolbar. The smaller window contained inside the main frame window labeled <untitled> is a new, blank Model Window. We will type our sample model directly into this window.

The following are the Syntax of LINDO:

(i) MAX or MIN will be used to represent the Maximization or Minimization objective function

(ii) The maximum allowable length of variable names is 8 characters

(iii) Constraints should be started after ST

(iv) Constraint Name should be terminated with a parenthesis

(v) (+, -, >, <, =) are the recognized arithmetic operations

(vi) Comments can be added by starting with an exclamation mark (!)

(vii) Splitting lines in a model is permitted in LINDO

(viii) LINDO is not case sensitive

(ix) RHS should contain only values and LHS should contain variables along with their coefficients

(x) Parentheses are not recognized in LINDO

Example A.8. *Use LINDO to solve Example A.1.*

Enter the problem as shown:

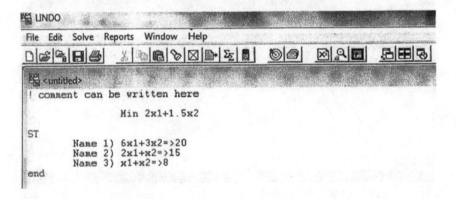

- Comments can be inserted starting with! (exclamation point).

- The constraint name can be given by terminating the name with a parenthesis)

Now, click "Solve" on the menu bar and then select "Solve." Click on "No" in the "Do Range (Sensitivity) Analysis?" dialog box. The solution will be displayed in a separate "Reports Window." The output is:

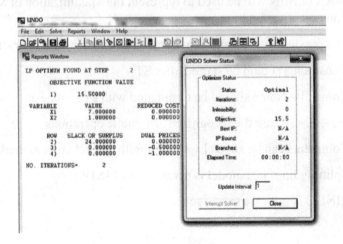

One can also see the final simplex tableau of the problem by going to "Report" on the menu bar and selecting the "Tableau" option and we have the following:

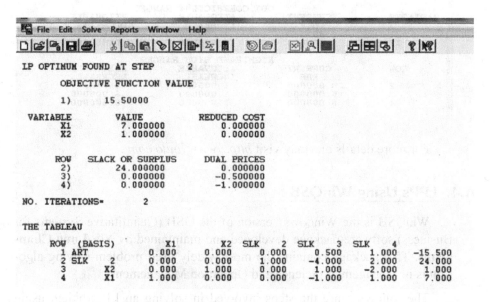

Also we can have the sensitivity analysis of the same problem by clicking on "Yes" in the "Do Range (Sensitivity) Analysis?" dialog box.

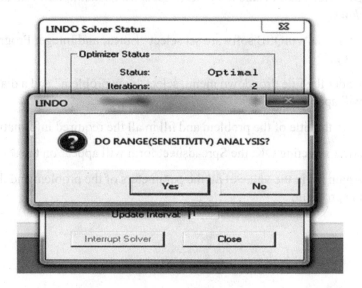

The following is the output of the sensitivity analysis:

```
RANGES IN WHICH THE BASIS IS UNCHANGED:

                            OBJ COEFFICIENT RANGES
VARIABLE          CURRENT         ALLOWABLE         ALLOWABLE
                   COEF           INCREASE          DECREASE
       X1        2.000000        1.000000          0.500000
       X2        1.500000        0.500000          0.500000

                            RIGHTHAND SIDE RANGES
   ROW            CURRENT         ALLOWABLE         ALLOWABLE
                    RHS           INCREASE          DECREASE
       2         20.000000       24.000000         INFINITY
       3         15.000000        1.000000          6.000000
       4          8.000000        7.000000          0.500000
```

For more details one may visit *http://www.lindo.com/*.

A.4. LPPs Using WinQSB

WinQSB is the Windows version of the QSB (Quantitative Systems for Business) software package developed and maintained by **Yih-Long Changruns**. This package contains the most widely used problem-solving algorithms in Management Science and Operation Management.

The following are the steps involved in solving an LP problem using WinQSB:

(i) Download and install WinQSB (*https://winqsb.en.uptodown.com/windows*).

(ii) From the WinQSB software set select "Linear and Integer Programming."

(iii) Under the File dropdown menu, select "new problem" and a dialog box will appear.

(iv) Give the title of the problem and fill in all the required information.

(v) After selecting OK, the Spreadsheet form will appear on the screen.

(vi) Again fill in the value of all the parameters of the problem and then solve the problem.

Example A.9. *Use WinQSB to solve Example A.6.*

Solution: From the WinQSB software we select "Linear and Integer Programming" and under the File dropdown menu, select "new problem" and the following dialog box has appeared:

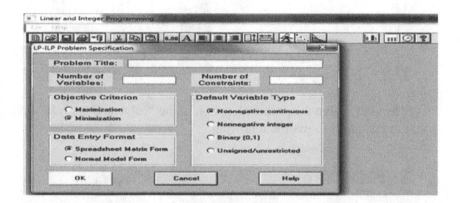

Fill in all the necessary information and click OK. A Spreadsheet form will appear on the screen, and the following is the screenshot of the Spreadsheet form after the values of the parameters are filled:

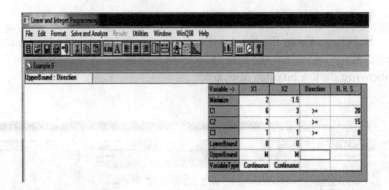

Select the "solve the problem" option in the "solve and analyze" option of the menu bar. It shows the following:

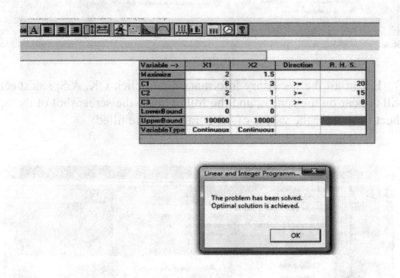

Click "ok" and get the following result:

19:51:28			Sunday	February	12	2017		
Decision Variable	Solution Value	Unit Cost or Profit c(i)	Total Contribution	Reduced Cost	Basis Status	Allowable Min. c(i)	Allowable Max. c(i)	
1 X1	100,000.0000	2.0000	200,000.0000	0	basic	0	M	
2 X2	10,000.0000	1.5000	15,000.0000	0	basic	0	M	
Objective	Function	(Max.) =	215,000.0000					

Constraint	Left Hand Side	Direction	Right Hand Side	Slack or Surplus	Shadow Price	Allowable Min. RHS	Allowable Max. RHS
1 C1	630,000.0000	>=	20.0000	629,980.0000	0	-M	630,000.0000
2 C2	210,000.0000	>=	15.0000	209,985.0000	0	-M	210,000.0000
3 C3	110,000.0000	>=	8.0000	109,992.0000	0	-M	110,000.0000

The final simplex table can be seen by selecting the "Final simplex Table" option from the "Result" option on the menu bar. We get the following output:

Basis	C(j)	X1 2.0000	X2 1.5000	Surplus_C1 0	Surplus_C2 0	Surplus_C3 0	Slack_UB_X1 0	Slack_UB_X2 0	Artificial_C1 0	Artificial_C2 0	Artificial_C3 0	R.H.S.	Ratio
X2	1.5000	0	1.0000	0	0	0	0	1.0000	0	0	0	10,000.0000	
Surplus_C3	0	0	0	0	0	1.0000	1.0000	1.0000	0	0	-1.0000	109,992.0000	
Surplus_C1	0	0	0	1.0000	0	0	6.0000	3.0000	-1.0000	0	0	629,980.0000	
X1	2.0000	1.0000	0	0	0	0	1.0000	0	0	0	0	100,000.0000	
Surplus_C2	0	0	0	0	1.0000	0	2.0000	1.0000	0	-1.0000	0	209,985.0000	
C(j)-Z(j)	0	0	0	0	0	-2.0000	-1.5000	0	0	0	215,000.0000		
* Big M	0	0	0	0	0	0	0	-1.0000	-1.0000	-1.0000	0		

REFERENCES

1. Antoine Augustin Cournot (1838), Researches into the Mathematical Principles of the Theory of Wealth, translated by Nathaniel T. Bacon, Macmillan, 1927.

2. Ê. Borel, La théorie du jeu et les équations intégrales à noyau symétrique gauche, C. R. Acad. Sci. Paris vol. 173 (1921), 1304 -1308.

3. Frederick S. Hillier and Gerald J. Lieberman, Introduction to Operations Research, McGraw Hill Education, 2011.

4. G. B. Dantzig and M. N. Thapa, Linear Programming 1: Introduction, Springer, 1997.

5. G. B. Dantzig, Origins of the Simplex Method, Stanford University: Systems Optimization Laboratory, 1987.

6. G. Hadley, Linear Programming, Addison Wesley, 1961.

7. G. Hadley, Linear Programming, Narosa Book Distributors, 2002.

8. Hamdy A. Taha, Operations Research, Pearson Education India, 2014.

9. John von Neumann and Oskar Morgenstern, Theory of Games and Economic Behavior, Princeton, University 1944.

10. Kanti Swarup, P. K. Gupta, and Man Mohan, Operations Research, Sultan Chand & Sons, 2014.

11. P. K. Gupta and Man Mohan, Linear Programming and Theory of Games, Sultan Chand & Sons, 2002.

12. Paul A. Jensen and Jonathan F. Bard, Operation Research Models and Methods, John Wiley & Sons, 2002.

13. Paul R. Thie and Gerard E. Keough, An Introduction to Linear Programming and Game Theory, John Wiley & Sons, 2008.

14. A. Ravindran D. T. Phillips and J. J. Solberg Operations Research—Principles and Practice, Second Edition, John Wiley & Sons, 1987.

15. Wayne L. Winston, Operations Research: Applications and Algorithms, Cengage Learning, 2003.

INDEX